T0257789

Encyclopedia of Alternative and Renewable Energy: Solar Cells and Dye Sensitized Devices

Volume 27

Encyclopedia of Alternative and Renewable Energy: Solar Cells and Dye Sensitized Devices

Volume 27

Edited by **Terence Maran and David McCartney**

New York

Published by Callisto Reference,
106 Park Avenue, Suite 200,
New York, NY 10016, USA
www.callistoreference.com

**Encyclopedia of Alternative and Renewable Energy:
Solar Cells and Dye Sensitized Devices
Volume 27**
Edited by Terence Maran and David McCartney

International Standard Book Number: 978-1-63239-201-5 (Hardback)

Contents

Permissions

List of Contributors

Preface

It is often said that books are a boon to mankind. They document every progress and pass on the knowledge from one generation to the other. They play a crucial role in our lives. Thus I was both excited and nervous while editing this book. I was pleased by the thought of being able to make a mark but I was also nervous to do it right because the future of students depends upon it. Hence, I took a few months to research further into the discipline, revise my knowledge and also explore some more aspects. Post this process, I begun with the editing of this book.

The book focuses on dye-sensitized solar cells (DSSCs). DSSCs are considered to be highly efficient as they are made of low cost materials with simple and economical manufacturing procedures and can be molded into malleable sheets. These cells have emerged as a new class of energy conversion tools, which represent the third generation of solar technology. The method of converting solar energy into electricity in these devices is quite distinctive. The energy conversion efficiency produced by the DSSCs is low, but, it has upgraded quickly in the last years. It is assumed that the DSSCs will take an admirable place in the large scale production for the future. This book covers various aspects related to DSSCs such as photon management strategies, metal substrates, carbon nanostructures as low cost and DSSCs as an alternative approach.

I thank my publisher with all my heart for considering me worthy of this unparalleled opportunity and for showing unwavering faith in my skills. I would also like to thank the editorial team who worked closely with me at every step and contributed immensely towards the successful completion of this book. Last but not the least, I wish to thank my friends and colleagues for their support.

Editor

Effective Methods for the High Efficiency Dye-Sensitized Solar Cells Based on the Metal Substrates

Ho-Gyeong Yun[1], Byeong-Soo Bae[2], Yongseok Jun[3] and Man Gu Kang[1]
[1]Convergence Components & Materials Research Lab., Electronics and Telecommunications Research Institute (ETRI), Daejeon,
[2]Lab. of Optical Materials and Coating (LOMC), Dep. of Materials Science and Eng. KAIST, Daejeon
[3]Interdisciplinary School of Green Energy, Ulsan National Institute of Science, Ulsan, Republic of Korea

1. Introduction

A nano porous dye-sensitized solar cell (DSSC) has been widely studied since its origin by O'Regan and Grätzel.[1] By virtue of many sincere attempts, a conversion efficiency of more than 11%[2] and long-term stability[3] has been achieved using a DSSC with F-doped SnO_2 layered glass (FTO-glass). However, relatively low conversion efficiency of the DSSC, compared with the crystalline Si (24.7%) or thin film CIGS (19.9%), restricts its further applications so far.[4] In order to improve the conversion efficiency of the DSSC, continuous attempts have been made in the past decades. Researchers have concentrated their attention on the working or counter electrode materials, synthesizing dye, additives of the electrolytes, nano-structures for enhancing light scattering and so on.[5-9] However, there have been few reports on the interface between nano-crystalline electrode material and current collecting substrates, in particular on the DSSC with thin and light-weight metal substrates. A DSSC with thin and lightweight substrate could extend its application. However, widely used conductive-layer-coated plastic films such as indium doped tin oxide (ITO) coated polyethylene terephthalate (PET) or polyethylene naphthalate (PEN) film degrade at the TiO_2 sintering temperature of approximately 500 ºC. Furthermore, thermal treatment of TiO_2 particles below plastic degeneration temperature causes poor necking of TiO_2 particles, resulting in a low conversion efficiency.[10] Several methods have been tried in order to answer to this problem, such as hydrothermal crystallization,[11] electrophoretic deposition under high DC fields,[12] and low temperature sintering.[13] However, these methods did not show the fundamental solution for the low necking problem. For better attempts, instead of plastic film, previous study has proposed thin metal foil as a substrates.[14-16] A thin metal foil can be a excellent alternative to conductive-layer-coated plastic films, because temperature limitation due to substrate could be eliminated.

Focusing on the characteristics of the interface between nano-sized TiO_2 and metal substrates, this chapter describes several effective methods for the high efficiency DSSCs based on metal substrates. Briefly, we report a increased light-to-electricity conversion efficiency and decreased electrical resistance of DSSC with the roughened StSt substrate.[17] In addition, an acid treatment of the Ti substrates for nanocrystalline TiO_2 photo-electrode prior to thermal oxidation significantly improved the optical and electrochemical behaviors at the same time, resulting in a highly increased performance in terms of all performance factors, i.e. V_{oc}, J_{sc}, FF, and efficiency.[18] Finally, a synergistic effect of vertically grown TiO_2 nano tube (TiO_2 NT) array and TiO_2 nano powder (TiO_2 NP) would also be introduced.[19] Detailed experimental procedures are not described in this chapter, because they are well explained in the references.

2. StSt and Ti substrates for photo-electrodes of the DSSCs

Considering the work function of the metals, promising metal substrates for DSSCs are Ti, StSt, tungsten (W) and Zinc (Zn)[14] because the work function determine the contact types, i.e. ohmic contact or schottky contact. In case of the n-type semiconductor such as TiO_2, the work function of the metal should be lower than that of semiconductor, ohmic contact. Furthermore, in the metals such as Ti, StSt, W, and Zn, the oxide layer produced by thermal treatment play important roles in the cell properties.[16] However, during thermal treatment, Al, Co, and etc generate insulating oxide layer, which make it insulator. Ti is most desirable metal substrate of the DSSCs because the thermally oxidized layer might have very similar structure with the nano-crystalline TiO_2 layer. The almost same electrochemical impedance of the W with the Ti was also reported. Under the assumption that most of the oxide layer is WO_3, the conduction band energy level of the W locates only 0.15 V below the one of TiO_2, as shown in Fig. 1[16] When the mutual disposition of energy levels is considered, the conduction band energy levels of the facing semiconductor metal oxides overlap.[20, 21] This overlapping does not significantly block the charge carriers flow, and no noticeable increase of the resistance has been reported.[16] However, W is not a common but rare metal. In the case of the StSt, some higher electrochemical impedance than Ti was reported due to conduction band energy level mismatch. However, StSt is most common and cost-effective material for the substrates of the DSSCs. Therefore, Ti and StSt are most frequently focused at the realization of the DSSCs on the metal substrates.[22-26]

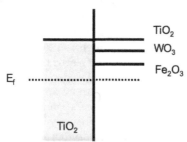

Fig. 1. Diagram of the conduction band edges of the semiconductor metal oxides. © The Electrochemical Society[16].

3. StSt substrate: effect of increased surface area[17]

The injection process used in the DSSC does not introduce a hole, i.e. minority carriers, in the TiO_2, only an extra electron.[27] On the contrary, as majority carriers and minority carriers, electrons and holes co-exist in *p-n* junction type solar cell, causing high electron/hole recombination rate. Therefore, in order to decrease the emitter recombination as much as possible, point-contact solar cells were introduced.[28, 29] In this paragraph, however, we report increased conversion efficiency and decreased electrical resistance of DSSCs with the roughened StSt substrates. Sulfuric acid-based solutions are effective StSt pickling reagents.[30] Additives, such as hydrated sodium thiosulphate and propargyl alcohol, endowed the StSt with pores and increased the surface area.[31] Under the atomic force microscope (AFM) analysis, the actual surface area of the roughened StSt substrates were measured to be a 23.6% increase. (Fig. 2)

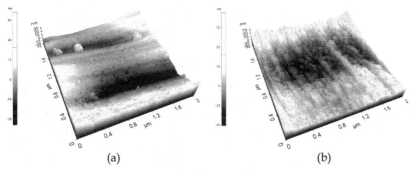

(a)　　　　　　　　　(b)

Fig. 2. AFM images of StSt surface (a) before and (b) after roughening process. © American Institute of Physics[17].

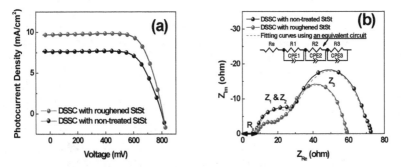

Fig. 3. Under AM 1.5 irradiation (100 mW/cm^2) with a xenon lamp. (a) J-V curves of DSSC with nontreated StSt substrates and roughened StSt substrates. (b) Electrochemical impedance spectra measured at the frequency range of 10^{-1}–10^6 Hz and fitting curves using an equivalent circuit model including three CPEs. © American Institute of Physics[17].

The J-V characteristics of the DSSCs with non-treated and roughened StSt substrates are shown in Fig. 3. (a). After roughening, the conversion efficiency and J_{sc} of the DSSC increased 33% and 27% respectively. However, open circuit voltage (V_{oc}) and fill factor (FF) remained nearly constant. V_{oc} changed from 800 mV to 807 mV and FF varied from 70.3% to 72.4% after roughening. To identify the cause of the increased J_{sc} and efficiency, electrochemical impedance spectra were measured in the frequency range of 10^{-1} to 10^6 Hz and the resistance from electrochemical impedance spectra was estimated using the equivalent circuit model including 3 constant phase elements (CPEs). (Fig. 3. (b)) Even though there were small differences in R_2 and R_3 after roughening, R_1 was reduced from 17.1 to 3.9. The largely reduced R_1 clearly comes from the reduced electrical resistance of the $TiO_2/StSt$ interface because R_1 represents the electrical resistance at this interface.[32] Considering the same electrical resistance between the TiO_2 particles and the interface with the Pt/electrolyte in DSSCs with both non-treated and roughened substrates, the small difference of R_2 after roughening is expected result. The value of R_3 is closely related to the reverse electron transfer from TiO_2 to the electrolyte.[32] In detail, as the number of electrons returning to the electrolyte increases, the arc of Z_3 increases. Therefore, the fact that R_3 remains unchanged after roughening clearly indicates that the increased electrical contact area does not cause an increase in reverse electron transfer.

4. Ti substrate: a simple surface treating method[18]

In this paragraph, we report that acid (HNO_3-HF) treatment of the titanium (Ti) substrate for the photo-electrode significantly improved the efficiency of DSSCs. Prior to spreading the TiO_2 paste, the Ti substrates were chemically treated with HNO_3-HF solution. As shown in Fig. 4 (a) and (b), HNO_3-HF treatment caused sharp steps at the grain boundaries, due to different etching rates of dissimilar crystal structures between the grains and the grain boundaries.[33] Fig. 5 (a) ~ (c) shows the cross-sectional scanning transmission electron microscopy (STEM) images of the Ti substrates. On the outermost surface, the non-treated Ti substrate exhibited a finer-grained structure. This suggests that the outermost surface of the Ti substrate was composed of finer-grained disordered Ti, which resulted from the thermo-mechanical manufacturing process.[34] However, treatment of the Ti substrate with the HNO_3-HF solution completely removed this finer-grained disordered region. Furthermore, the thermally oxidized layer of the non-treated substrate was much thicker and more variable than that of the HNO_3-HF-treated substrates. (Fig. 5 and 6) In the field emission transmission electron microscope (FE-TEM) analysis, the oxidized layer of the non-treated Ti substrate, which was produced by oxygen diffusion to the finer-grained disordered region, showed a disordered grain structure, i.e. a low degree of crystallinity. However, the oxide layer of HNO_3-HF-treated Ti substrates, which was developed by the oxygen diffusion into the normally-grained Ti substrate, was almost a single crystal. The corresponding X-ray diffraction (XRD) patterns also showed that the HNO_3-HF treatment had produced a variation on the phase and crystallinity of a thermally oxidized layer.

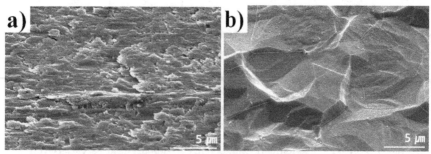

Fig. 4. SEM images of the Ti surface before thermal annealing: (a) non-treated, (b) HF-HNO3 treated. © WILEY-VCH Verlag GmbH & Co. KGaA, Weinheim[18].

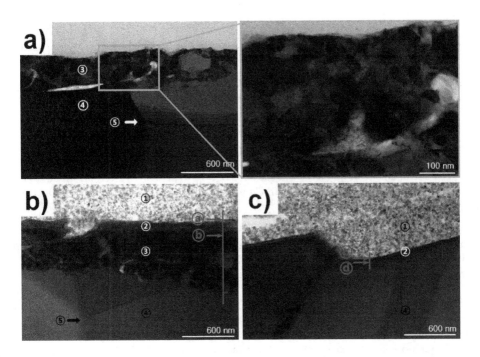

Fig. 5. Cross-sectional STEM images of Ti substrates (a) untreated substrate before thermal annealing, including a magnified view of the finer grained disordered region, (b) untreated substrate after thermal annealing at 550 ℃ for 30 min, (c) HF-HNO₃-treated substrate after thermal annealing. Note: ① sintered TiO_2 particles, ② thermally oxidized Ti, ③ finer-grained disordered Ti, ④ normally grained Ti, ⑤ normal grain-boundaries of Ti. © WILEY-VCH Verlag GmbH & Co. KGaA, Weinheim[18].

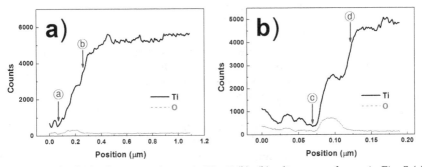

Fig. 6. EDX graph of (a) a line-scan shown in Fig. 5 (b), (b) a line-scan shown in Fig. 5 (c). © WILEY-VCH Verlag GmbH & Co. KGaA, Weinheim[18].

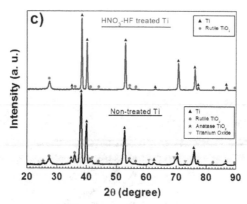

Fig. 7. By use of a 2θ scan method, XRD patterns of non-treated and HNO₃-HF-treated Ti substrates after thermal annealing at 550 °C for 30 min. © WILEY-VCH Verlag GmbH & Co. KGaA, Weinheim[18].

As shown in Fig. 7, the thermally oxidized layer of the non-treated Ti substrate exhibited various oxide forms including anatase TiO_2, rutile TiO_2, and titanium oxide. However, only rutile TiO_2 was observed in the oxide layer of the HNO₃-HF-treated Ti substrate.

The variation of the microstructures influenced the optical and electrochemical behaviour at the same time resulting in highly increased efficiency, 9.20%. (Fig. 8 (d)) Fig. 8 (a) shows the optical reflection of the Ti. The HNO₃-HF-treated substrate exhibited a significantly increased optical reflection. The low and flat reflection behavior of the non-treated Ti substrates could be attributed to the thick and non-uniform thickness of the oxide layer and the inferior optical reflectance at the inner metal surface.[35, 36] In the evaluation of the illumination intensity effect on the performance factors, the V_{oc} and J_{sc} exhibited logarithmic and linear dependence respectively. However, FF decreased under stronger illumination intensity. These consequences suggest that the improved performance of the DSSC with the HNO₃-HF-treated substrate cannot be attributed to the enhanced optical reflection alone. Rather, the greater part of this improvement could be attributed to a reduced back reaction of the electrons with I_3^- ions at the interface of the conductive substrate and electrolyte because the thickness of the nano crystalline TiO_2 layer is about 15μm. For a device with a > 10 μm thick TiO_2 layer, performance increases due to reflection are restricted to wavelengths above 580 nm where the absorption of the N719 dye is weak.[15]

The blocking layer (compact TiO_2) at the interface of the TiO_2 particles/conductive substrates has been studied[37, 38] and several groups concluded that recombination occurs predominantly near the conductive substrate and not across the entire TiO_2 film.[39] In the DSSCs with metal substrates, the oxidized layer is naturally formed at the interface of the TiO_2 particles/conductive substrate during thermal annealing. However, it seems that the low quality oxidized layer induced poor blocking behavior of the DSSCs with the non-treated Ti substrates. The recombination kinetics were investigated by the evaluation of the rate of photovoltage decay. The rate of photovoltage decay is inversely proportional to the lifetime of the photoelectron in the DSSCs, and the lifetime of the electron is inversely proportional to the rate of recombination.[40] The HNO₃-HF treatment of the Ti substrates strongly influenced the rate of the photovoltage decay. (Fig. 8 (b)) The electron recombination may lead to a lowering of the photocurrent, but also to a decrease in the photovoltage by lowering the quasi-Fermi level for the electrons under illumination due to a kinetic argument.[41, 42] Furthermore, the *FF* is a measure of the increase in recombination (decrease in photocurrent) with increasing photovoltage.[43] If the improved optical reflection at the substrate were a dominant element of enhanced performance, the V_{oc} and *FF* would restrictively increase and decrease respectively. An obviously possible cause for the significantly improved performance is decreased recombination at the interface of the TiO_2/conductive substrate after HNO_3-HF treatment.

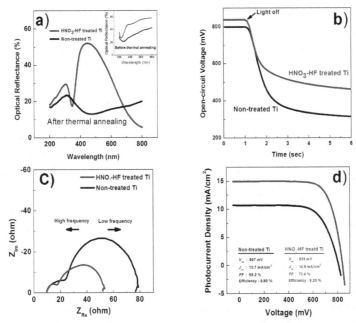

Fig. 8. (a) Optical reflectance of Ti substrates measured with UV-VIS-NIR spectro-photometers combined with an integrated sphere before and after thermal annealing at 550 °C for 30 min. Baseline calibration was performed with a standard specimen composed of polytetrafluoroethylene (PTFE). (b) open-circuit voltage decay measurement, (c) electrochemical impedance spectra, and (d) J-V curves of DSSC with non-treated and HNO_3-HF-treated Ti substrates. © WILEY-VCH Verlag GmbH & Co. KGaA, Weinheim[18].

As shown in Fig. 8 (c), electrochemical impedance also improved after HNO_3-HF treatment. The 1st semicircle is closely related to charge transfer at the counter electrode and/or electrical contact between conductive substrate/TiO_2 or TiO_2 particles.[22] The relatively small size of the 1st semicircles (high frequency range) in the cell with the HNO_3-HF-treated substrate indicated a reduced electrical resistance, i.e. improved contact at the TiO_2/Ti interface accordingly. Furthermore, the size of the 2nd semicircle (low frequency range) was also largely decreased. The 2nd semicircle is related to the recombination of electrons with I_3^- .[14] Under the assumption that the micro-structures of the oxidized layers determine the blocking ability, the significantly decreased size of the 2nd semicircle could be attributed to a highly decreased charge recombination by virtue of improved micro-structure after HNO_3-HF treatment of the Ti substrate.

5. Hybrid substrate: TiO$_2$ NP on the TiO$_2$ NT grown Ti substrates[19]

In the case of DSSCs based on metal substrates, light illumination should come from a counter electrode, i.e., back illumination. Therefore, the light scattering layer,[9] which enhances the optical path length, should be located between 20 nm sized TiO$_2$ nano-particles (NPs) and conductive substrates. This structure causes poor adhesion due to the large particle size of the scattering layer. Considering slow recombination and light scattering,[44] TiO$_2$ nano-particles has been incorporated on the short TiO$_2$ nano-tube grown Ti substrates. The preparation of photo-electrode is completed by four steps: ① anodization of a Ti foil for the formation of short TiO$_2$ NT arrays, ② doctor blading of TiO$_2$ NP included paste on the TiO$_2$ NT formed Ti substrates, ③ thermal treatment of the photo-electrode prepared by step ① & ②, ④ dye coating. In our approach, therefore, the fabrication time and length of TiO$_2$ NT could be minimized without diminishing, rather increasing, the surface area of the photo-electrode.

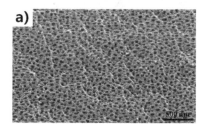 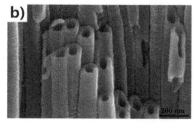

Fig. 9. SEM images of the TiO$_2$ NT fabricated Ti for 30 min anodizing (a) surface and (b) cross-section. © The Royal Society of Chemistry[19].

Out of several fabrication method of the TiO$_2$ NT such as electrochemical anodizing,[45] hydrothermal synthesis,[46] and template-assisted synthesis,[47] anodizing is a relatively simple approach for the preparation of optimized TiO$_2$ NT.[48] Anodizing at 50 V in a solution of ethylene glycol containing ammonium fluoride (NH$_4$F) resulted in the formation of regular TiO$_2$ NT arrays. (Fig. 9) When the anodizing was performed for 30 min, the tube diameter and wall thickness were estimated to be about 100 and < 50 nm, respectively. The lengths of the TiO$_2$ NT layers were controlled by the anodizing time. When the anodizing was performed for 15, 30, and 60 min, the lengths of the TiO$_2$ NTs were 1.53, 4.36, and 8.17 μm, respectively. TiO$_2$ NT and TiO$_2$ NP bonded well following thermal annealing at 550 °C for 30 min. (Fig. 10)

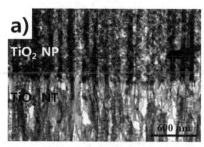 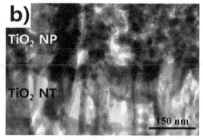

Fig. 10. Cross-sectional TEM images of (a) interface between TiO$_2$ NP and TiO$_2$ NT (b) magnified view of (a). © The Royal Society of Chemistry[19].

As is same with the previous paragraph, the TiO$_2$ NP film was made 15 μm thick, because that was the size that allowed DSSCs to exhibit optimal performance. When the thickness of the TiO$_2$ NP was more than 15 μm, the DSSC with TiO$_2$ NP on the Ti substrate (TiO$_2$ NP/Ti) exhibited a lowered performance because the thick TiO$_2$ layer (> 15 μm) provided additional electron recombination sites, resulting in a decreased open-circuit voltage (V_{oc}) and fill factor (*FF*).[49] However, a performance of the DSSCs with TiO$_2$ NP+NT/Ti increased continuously with increasing TiO$_2$ NT thickness up to 30 min anodized TiO$_2$ NT. (Fig. 11 (a)) This difference between the DSSC with TiO$_2$ NP+NT/Ti and the DSSC with TiO$_2$ NP/Ti can be attributed to the TiO$_2$ NT having an electron recombination that was reduced by comparison with the TiO$_2$ NP. The electron lifetime in the TiO$_2$ NT was longer than that in the TiO$_2$ NP because of the electron-recombination suppression from the reduction in electron-hopping across the inter-crystalline contacts between the grain boundaries.[50] As is described in the previous paragraph, optical transmission is restricted to wavelengths > 570 nm for a device with a TiO$_2$ layer that is more than 10 μm thick, resulting in a restricted increase in J_{sc}.[15] However, strong internal light scattering within the TiO$_2$ NTs elongated the path length of the long-wavelength incident light to promote the capture of photons by the dye molecules.[44] Despite a surface area of the DSSC with TiO$_2$ NP on 30-min-anodized TiO$_2$ NT/Ti that was smaller than that of DSSC with 20 μm thick TiO$_2$ NP/Ti, the increased J_{sc} could also be a result of stronger light scattering effects.

The reduced electron recombination at the interface of the TiO$_2$ NT/electrolyte was also represented in an electrochemical impedance measurement (Fig. 11 (b)). Under the assumption that the TiO$_2$ NT is superior to TiO$_2$ NP in the interfacial contact with Ti substrates due to the *in-situ* fabrication process, the largely reduced size of the 1st semicircle in a DSSC with TiO$_2$ NP+NT/Ti could be a result of the reduced electrical resistance at the interfacial contact. However, the size of the 2nd semicircle (low frequency range) was almost the same. The 2nd semicircle represents the recombination of injected electrons to the TiO$_2$ film with electrolyte.[51] Furthermore, the DSSCs with TiO$_2$ NP/Ti and TiO$_2$ NP+NT/Ti exhibited a similar rate of photovoltage decay, which is proportional to the rate of recombination (Fig. 11 (c)). The overall TiO$_2$ film in the DSSC with TiO$_2$ NP+NT/Ti was thicker than that of the DSSC with TiO$_2$ NP/Ti due to the introduction of the TiO$_2$ NT layer at the interface of the TiO$_2$ NP and Ti substrate. Therefore, it seems that the small variation in the 2nd semicircle in the electrochemical impedance spectra and the rate of photovoltage decay can be attributed to the slow recombination characteristics of the TiO$_2$ NT.

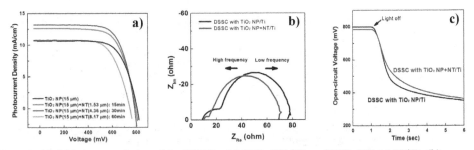

Fig. 11. (a) *J–V* characteristics of the DSSC with TiO$_2$ NP/Ti and TiO$_2$ NP + NT/Ti. (b) Electrochemical impedance spectra in frequencies ranging from 10^{-1} to 10^6 Hz. (c) Open-circuit voltage decay measurement. © The Royal Society of Chemistry[19].

6. Conclusion

Several methods for the high efficiency DSSCs based on the metal substrates have been introduced. In the case of the StSt substrate, the solar cell performance was significantly improved by the roughening process, which enhances electrical contact by roughening the substrates. In addition, when a Ti substrate was treated with an acid solution, both the surface morphology and the crystalline structure of the thermally oxidized layer were varied, resulting in the simultaneous improvements in V_{oc}, J_{sc} and FF. Finally, the DSSCs with TiO_2 NP + NT/Ti were prepared for the synergistic effect of vertically grown TiO_2 NT and TiO_2 NP films. The slow electron recombination at the interface of the TiO_2 NT/electrolyte and the light scattering effect might have simultaneously contributed to DSSC performance, resulting in the improved J_{sc} and conversion efficiency with only a negligible effect on the V_{oc} and FF.

7. Acknowledgements

This article is prepared and reproduced under the permission of the American Institute of Physics, WILEY-VCH Verlag GmbH & Co. KGaA, The Electrochemical Society, and The Royal Society of Chemistry. Each article has been referred at the corresponding section.

8. References

[1] B. O'Regan, M. Grätzel, Nature 353, 737 (1991)
[2] M. K. Nazeerudden, P. Pechy, T. Renouard, S. M. Zakeeruddin, B. R. Humphry, P. Comte, P. Liska, L. Cevey, E. Costa, V. Shklover, L. Spiccia, G. B. Deacon, C. A. Bignozzi, M. Grätzel, J. Am. Chem. Soc. 123, 1613 (2001)
[3] P. Wang, S. M. Zakeeruddin, J. E. Moser, K. Nazeeruddin, T. Sekiguchi, M. Grätzel, Nat. Mat. 2, 402 (2003)
[4] M. A. Green, K. Emery, Y. Hishikawa, W. Warta, Progress in Photovoltaics 17, 320 (2009)
[5] Y. J. Kim, M. H. Lee, H. J. Kim, G. Lim, Y. S. Choi, N. -G. Park, K. Kim, W. I. Lee, Adv. Mater. 21, 3668 (2009)
[6] A. Nattestad, A. J. Mozer, M. K. R. Fischer, Y. –B. Cheng, A. Mishra, P. Bäuerle, U. Bach, Nature Mater. 9, 31 (2010)
[7] F. Gao, Y. Wang, J. Zhang, D. Shi, M. Wang, R. Humphry-Baker, P. Wang, S. M. Zakeerudding, M. Grätzel, Chem. Commun. 2635 (2008)
[8] M. Wang, X. Li, H. Lin, P. Pechy, S. M. Zakeeruddin, M. Grätzel, Dalton Trans. 10015 (2009)
[9] S. Ito, S. M. Zakeeruddin, R. Humphry-Baker, P. Liska, R. Charvet, P. Comte, M. K. Nazeeruddin, P. Péchy, M. Takata, H. Miura, S. Uchida, M. Grätzel, Adv. Mater. 18, 1202 (2006)
[10] C. Y. Jiang, X. W. Sun, K. W. Tan, G. Q. Lo, A. K. K. Kyaw, D. L. Kwong, Appl. Phys. Lett. 92, 143101 (2008)
[11] D. Zhang, T. Yoshida, H. Minoura, Adv. Mater. 15, 814 (2003)
[12] D. Mattews, A. Kay, M. Grätzel, Aust. J. Chem. 47, 1869 (1994)
[13] C. Longo, A. F. Nogueira, M. A. De Paoli, H. Cachet, J. Phys. Chem. B 106, 5925 (2002)
[14] M. G. Kang, N. G. Park, K. S. Ryu, S. H. Chang, K. J. Kim, Sol. Energy Mater.Sol. Cells 90, 574 (2006)

[15] Y. Jun, J. Kim, M. G. Kang, *Sol. Energy Mater. Sol. Cells* 91, 779 (2007)
[16] Y. Jun, M. G. Kang, *J. Electrochem. Soc.* 154, B68 (2007)
[17] H. -G. Yun , Y. Jun , J. Kim , B. -S. Bae , M. G. Kang , *Appl. Phy. Lett.* 93, 133311 (2008)
[18] H. -G. Yun , B. -S. Bae, M. G. Kang , *Advanced Energy Materials* 1, 1 (2011)
[19] H. -G. Yun , J. H. Park , B. -S. Bae , M. G. Kang , *J. Mater. Chem.* 21, 3558 (2011)
[20] M. K. Kang, N. G. Park, S. R. Kwang, H. C. Soon, K. J. Kim, *Chem. Lett.* 34, 804 (2005)
[21] H. H. Kung, H. S. Jarrett, A. W. Sleight, A. Ferretti, *J. Appl. Phys.* 48, 2463 (1977)
[22] J. H. Park, Y. Jun, H. –G. Yun, S. –Y. Lee, M. G. Kang, *J. of Electrochem. Soc.* 155, F145 (2008)
[23] H. Lindström, A. Holmberg, E. Magnusson, S. Lindquist, L. Malmqvist, A. Hagfeldt, *Nano Lett.* 1, 97 (2001)
[24] S. Uchida, M. Tomiha, H. Takizawa, M. Kawaraya, *J. of Photochem. and Photobio. A: Chem* 164, 93 (2004)
[25] T. Miyasaka, Y. Kijitori, *J. of Electrochem. Soc.* 151, A1767 (2004)
[26] M. Dürr, A. Schmid, M. Obermaier, S. Rosselli, A. Yasuda, G. Nells, *Nature Mat.* 4, 607 (2005)
[27] J. N. Hart, Y. -B. Cheng, G. P. Simon, L. Spiccia, *J. of Nanoscience and Nanotech.* 8, 2230 (2008)
[28] T. Markvat, L. Castaner, Solar Cells: Materials, Manufacture and Operation, Elsevier Science, Oxford, 377 (2005)
[29] R. M. Swanson, S. K. Beckwith, R. A. Crane, W. D. Eaides, Y. O. Kwark, R. A. Sinton, S. E. Swiiwiun, *IEEE Trans. on Elec. Dev.* 31, 5 (1984)
[30] A. Tamba, N. Azzerri, *J. App. Electrochem.* 2, 175 (1972)
[31] S. E. Hajjaji, M. E. Alaoui, P. Simon, A. Guenbour, A. Ben Bachir, E. Puech-Costes, M. T. Maurette, L. Aries, *Sci. and Tech. of Adv. Mat.* 6, 519 (2005)
[32] T. Hoshikawa, M. Yamada, R. Kikuchi, K. Eguchi, *J. Electrochem. Soc.* 152, E68 (2005)
[33] J. Lichtscheidl, K. J. Hartig, N. Getoff, C. Tauschnitz, G. Nauer, *Z. Naturforsch. A* 36a , 727 (1981)
[34] P. R. F. Barnes, L. K. Randeniya, P. F. Vohralik, I. C. Plumb, *J. Electrochem. Soc.* 154 , H249 (2007)
[35] G. Jerkiewicz, H. Strzelecki, *Langmuir* 12, 1005 (1996)
[36] J. –L. Delplancke, M. Degrez, A. Fontana, R. Winand, *Surface Tech.* 16 , 153 (1982)
[37] J. Xia, N. Masaki, K. Jiang, S. Yanagida, *Chem. Commun.* 138 (2007)
[38] B. Peng, G. Jungmann, C. Jäger, D. Haarer, H.-W. Schmidt, M. Thelakkat, *Coord. Chem. Rev.* 248, 1479 (2004)
[39] K. Zhu, E. A. Schiff, N. -G. Park, J. V. D. Lagemaat, A. J. Frank, *Appl. Phys. Lett.* 80, 685 (2002)
[40] A. Zaban, M. Greenshtein, J. Bisquetr, *Chem. Phys. Chem* 4, 859 (2003)
[41] Hagfeldt, M. Grätzel, *Chem, Rev.* 95, 49 (1995)
[42] Kumar, P. G. Santangelo, N. S. Lewis, *J. Phys. Chem.* 96, 834 (1992)
[43] D. Cahen, G. Hodes, M. Grätzel, J. F. Guillemoles, I. Riess, *J. Phys. Chem. B* 104, 2053 (2000)
[44] K. Zhu, N. R. Neale, A. Miedaner, A. J. Frank, *Nano Lett.* 7, 69 (2007)
[45] D. Gong, C. A. Grimes, O. K. Varghese, W. C. Hu, R. S. Singh, Z. Chen, E. C. Dickey, *J. Mater. Res.* 16, 3331 (2001)
[46] T. Kasuga, M. Hiramatsu, A. Hoson, T. Sekino, K. Niihara, *Langmuir* 14, 3160 (1998)

[47] P. Hoyer, *Langmuir* 12, 1411 (2006)
[48] G. K. Mor, K. Shankar, M. Paulose, O. K. Varghese, C. A. Grimes, *Nano Lett.* 5, 191 (2005)
[49] R. Kato, A. Furube, A. V. Barzykin, H. Arakawa, M. Tachiya, *Coord. Chem. Rev.* 248, 1195 (2004)
[50] C.-J. Lin, W.-Y. Yu, S.-H. Chien, *Appl. Phys. Lett.* 93, 133107 (2008)
[51] T. Hoshikawa, M. Yamada, R. Kikuchi, K. Eguchi, J. Electrochem. Soc. 152, E68 (2005)

Dye Solar Cells: Basic and Photon Management Strategies

Lorenzo Dominici[1,2] et al.[*]
[1]Centre for Hybrid and Organic Solar Energy Centre (CHOSE), Dept. of Electronic Eng.,
Tor Vergata University of Rome, Roma
[2]Molecular Photonics Laboratory, Dept. of Basic and Applied Physics for Eng.,
SAPIENZA University of Rome, Roma
Italy

1. Introduction

After the introduction in 1991 by B. O'Regan and M. Grätzel, Dye Solar Cells (DSCs) have reached power conversion efficiency values over small area device as high as 11%. Being manufactured with relatively easy fabrication processes often borrowed from the printing industry and utilizing low cost materials, DSC technology can be considered nowadays a proper candidate for a large scale production in industrial environment for commercial purposes.

This scenario passes through some challenging issues which need to be addressed such as the set-up of a reliable, highly automated and cost-effective production line and the increase of large area panels performances, in terms of efficiency, stability and life-time of the devices.

In this work, an overview of the most utilized DSCs materials and fabrication techniques are highlighted, and some of the most significant characterization methods are described. In this direction, different approaches used to improves devices performances are presented.

In particular, several methods and techniques known as Light Management (LM) have been reported, based to the ability of the light to be confined most of the time in the cell structure. This behavior contributes to stimulate higher levels of charge generation by exploiting scattering and reflection effects.

The use of diffusive scattering layers (SLs), nanovoids, photonic crystals (PCs), or photoanodes co-sensitization approaches consisting in the use of two Dyes absorbing in two different parts of the visible range, have been demonstrated to be effective strategies to carry out the highest values of device electrical parameters.

Finally, to increase the light path inside the DSCs active layer, the use of refractive element on the topside (a complementary approach to SL) has been shown a promising possibility to further improve the generated photocurrent.

[*]Daniele Colonna[1], Daniele D'Ercole[1], Girolamo Mincuzzi[1], Riccardo Riccitelli[3], Francesco Michelotti[2], Thomas M. Brown[1], Andrea Reale[1] and Aldo Di Carlo[1]
1 Centre for Hybrid and Organic Solar Energy Centre (CHOSE), Dept. of Electronic Eng., Tor Vergata University of Rome, Roma, Italy
2 Molecular Photonics Laboratory, Dept. of Basic and Applied Physics for Eng.,
SAPIENZA University of Rome, Roma, Italy
3 DYERS srl, Roma, Italy

2. Material and processing for dye solar cell technology

Since the introduction and development of the dye-sensitized solar cell (DSC) (O'Regan & Graetzel, 1991) several efforts have been made to optimize the materials involved in the photo-electrochemical process and to improve the light conversion efficiency of the device (Hagfeldt & Graetzel, 1995), by exploiting a low cost production process based on simple fabrication methods, similar to those used in printing processes.

2.1 Dye solar cells architecture and working principle

In Fig. 1 the basic configuration of a Dye Solar Cell (DSC) is sketched (Chappel et al., 2005). Amongst the main elements of this electrochemical photovoltaic device is a mesoporous nanocrystalline Titanium Dioxide (nc-TiO_2) film deposited over a transparent and conductive layer coated glass (in particular Soda Lime or Borosilicate). As alternative to the nc-TiO_2, other large band-gap semiconductors (such as ZnO, Nb_2O_5, and $SrTiO_3$) can be utilized as well as flexible and plastic (PET or PEN) substrates are a valid options in substitution of the glass. Generally, the nc-TiO_2 film thickness is fixed to a value comprised between few microns to few tens of microns and the substrates conductivity is provided by a transparent conducting oxide (TCO) coating. Fluorine doped tin oxide (SnO_2:F), FTO is the most commonly used (although Tin doped Indium Oxide (In_2O_3:SnO_2) ITO is frequently found onto plastic substrates) since enables low cost massive production of substrates showing a sheet resistance as low as 8 Ωcm^{-2} (Solaronix®).

On the surface of the TiO_2, a monolayer of visible light harvesting dye molecules is adsorbed resulting in the TiO_2 visible light sensitizing. A wide variety of dye molecules, included naturals dyes extracted from fruits flowers or leaves, have been proposed and tested (Polo et al., 2004; Kalyanasundaram & Graetzel, 1998). At the moment metallorganic ruthenium complexes containing anchoring groups such as carboxylic acid, dihydroxy, and phosphonic acid on pyridyl ligands show the best performances. Largely diffused are in particular dyes commonly named as N3, black dye, N719 and Z907, which enable fabrication of highly performing devices (Kroon et al., 2007; Nazeeruddin et al., 2005; Z. S. Wang et al., 2005).

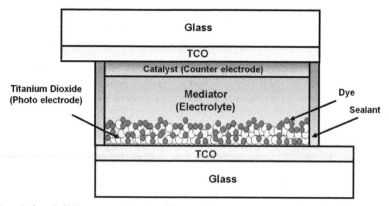

Fig. 1. Dye Solar Cell Structure. Basic cell's constituent are a transparent conductive substrate (TCO) coated glass and over it a nc-TiO_2 layer sensitized by a monolayer of adsorbed dye (photo-electrode), a red-ox mediator and, finally a catalyst (Pt) coated conductive substrate (counter-electrode).

It is worth to point out that the nc-TiO$_2$ mesoporous morphology (Fig. 2), for a film thickness of 10 μm, leads to an effective surface area about 1000 times larger as compared to a bulk TiO$_2$ layer, allowing for a significantly large number of sites offered to the dye sensitizer (Chen & Mao, 2007).

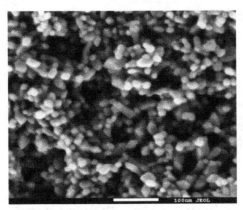

Fig. 2. A SEM image of nc-TiO$_2$ film utilized for Dye Solar Cells fabrication is shown. Although is possible to distinguish each nanoparticles (with a diameter of around 20 nm) large aggregates are evident resulting in a characteristic meso-porous morphology (Mincuzzi et al., 2011).

The conductive substrate together with the dye sensitized film form the cell photo-electrode. The dye sensitized film is placed in contact with a red-ox mediator electrolyte or an organic hole conductor material. The former is obtained solving a red-ox couple (such for instance the ions I$_3^-$/ I$^-$ or Co$^{(III)}$/Co$^{(II)}$) in a solvent such as 3-me-thoxypropionitrile (MPN), acetonitrile (ACN) or valeronitrile (VN). Finally the device is completed with a counter-electrode generally composed of a transparent and conductive substrate on which a Pt film of few tens of nanometers is deposited for red-ox catalysis purpose. The two electrodes and the red-ox electrolyte mediator are sealed together, evoking the characteristic picture of a sandwich-like structure. The most diffused sealants are thermoplastic gaskets typically made of Surlyn® (i.e. random copolymer poly(ethylene-co-methacrylic acid) – EMAA), Bynel® (modified ethylene acrylate polymer) or alternatively vitreous pastes or glass frits. Additionally, an optional light scattering layer made of particles with diameter of few hundreds of nanometers may be applied on top of the TiO$_2$ film in order to increase the photons optical path (transmitted light will be in fact back scattered into the nc-TiO$_2$ layer) and therefore the light harvesting (Hore et al., 2006). Although it has been demonstrated that this plays a beneficial role, a scattering layer leads to an opaque (not transparent) DSC device preventing the possibility of the use of the panel for building integration (as an active window for instance), one of the most interesting features and applications of DSC technology.

The working principle of DSC can be readily explained in terms of the electrons kinetics process and electrons transfer reactions taking place into the cell as a consequence of photon absorption. Fig. 3 shows the energy diagram and electrons transfer paths involved in a DSC.

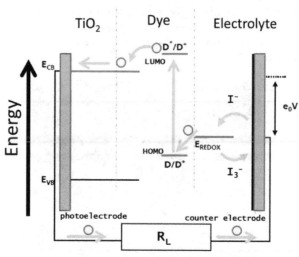

Fig. 3. DSC working principle: the absorption of a photon by a Dye molecule in its ground state D induce the transition to the excited state D*. The injection of an e- into the TiO$_2$ conduction band occurs, resulting in the Dye oxidation D+. The e- diffuse into the TiO$_2$ reaching an external circuit and a load R$_L$ where electrical power is produced. Successively it is reintroduced into the cell by the counter electrodes and regenerate the oxidized Dye D+ utilizing a redox couple as mediator.

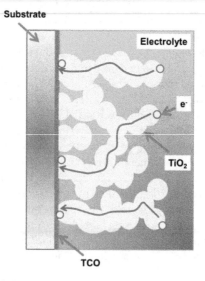

Fig. 4. The diffusion paths of the injected photon into the nc-TiO$_2$ conduction band are sketched. The nc-TiO$_2$ film sintering process promotes the electromechanical bonding between the nanoparticles facilitating the charge diffusion process in terms of increment of the intra-particles hopping rate.

By absorption of a photon, a dye molecule is set to the excited state D* from its ground state D. It follows that an electron is promoted from the highest occupied molecular orbital (HOMO) to the lowest unoccupied molecular orbital (LUMO) and then rapidly injected into the conduction band (E_{CB}) of the TiO_2. The dye molecule is then oxidized (D^+) whilst an hole is injected into the electrolyte. Charge transport occurs in the conduction band of the TiO_2 by pure diffusion of electrons percolating through the interconnected nc-TiO_2 particles to the FTO electrode (see Fig. 4). No electronic drift has been detected and electric fields in the TiO_2 are screened by the cations in the electrolyte, which penetrate the nano-scale pores of the TiO_2 (Van de Lagemaat et al., 2000).

Upon reaching the TCO electrode, the electrons are conducted to the counter-electrode via the external load (R_L) generating electrical power. Catalyzed by the platinum on the counter-electrode, the electrons are accepted by the electrolyte. This means, that the holes in the electrolyte (the I_3^-) recombine with electrons to form the negative charge carriers,

$$I_3^- + 2e- \rightarrow 3I^-$$

By diffusion, the negative charge (I^-) is transported back with the aim to reduce the oxidized dye molecule (D^+). Triiodide (I_3^-) is formed and the electrical circuit is closed:

$$2D^+ + 3I^- \rightarrow I_3^- + 2D$$

Therefore, in DSC device charge separation and charge transport occur in different media spatially separated. Apparently the device generates electric power from light without suffering any permanent chemical transformation. Nevertheless, going in deep toward a device realistic and detailed modeling, processes involving current losses have to be included and remarked.

When a photon is absorbed, a dye molecule is set in an excited state S*, the back relaxation (or back reaction) into its ground state S may also occur, preventing the injection of an electron into the TiO_2 (see Fig. 5). As shown by Nazeeruddin et al. (Nazeeruddin et al., 1993) the injection process has a much larger probability to occur resulting in a characteristic injection time in the range of fs -ps, which is more than 1000 times faster than back relaxation (about 10 ns) straightforwardly negligible.

As reported above, electrons injected into the conduction band of the TiO_2 diffuse through the TiO_2 film toward the FTO anode contact. The electron transport into the mesoporous nc-TiO_2 film can be modelled as a combination of two mechanisms electron hopping between sites and multiple trapping/detrapping (Bisquert et al., 2009). In the latter case, electrons spend part of their time immobilized in trap sites from which they are excited thermally back to the conduction band. Nevertheless, during their transit, there is a significant probability that an electron recombine (and be lost) with the oxidized dye molecule S^+, before the dye reduction caused by the electrolyte. We are facing, nonetheless, to a process with a characteristic time of several hundreds of nanoseconds resulting 100 times slower than the reduction induced by the electrolyte (~10 ns) (Hagfeldt & Graetzel, 1995).

Instead, electrons injected into the TiO_2 conduction band may, during the diffusion, more often recombine with the holes in the electrolyte, i.e. I_3^-. This constitutes the most significant electron loss mechanism in the DSC and it can be asserted that the electrons transport by diffusion in the nc-TiO_2, and their recombination with the electrolyte are the two competing processes in the DSC technology, affecting the device efficiency of electrons collection (Peter & Wijayantha, 2000). It is important to point out that, although the triiodide concentration in a DSC should be small for this reason, it should be high enough as to provide right amount of recombination for the electrons at the Pt counter-electrode. If this is not the case, the maximum current of the DSC will be diffusion-limited, i.e. cut by the diffusion of triiodide.

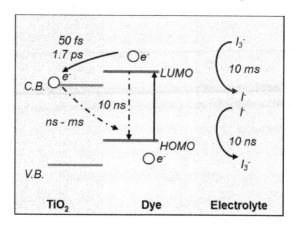

Fig. 5. Electron losses processes taking place into a DSC are shown. Also shown are the characteristic time relative to each of the considered processes.

Finally, it should be noted, that TiO_2 is a semiconductor with a large band gap of 3.2 eV, corresponding to a wavelength of λ=390 nm. Accordingly, visible light is not absorbed by the TiO_2 film. Direct absorption by UV-light is unwanted, since the created holes in the valence band of the TiO_2 are highly reactive and tends to produce the so called side reactions in the mediators, highly destructive for the long life time of the cell (Hinsch et al., 2001).

2.2 Dye solar cells fabrication

One of the main advantages connected with the DSC technology is the possibility to easily implement the fabrication steps involved, often borrowed from the printing industry, processing abundant materials with a relative low mass production costs (Di Carlo et al., 2008).

Differently from others photovoltaic technologies (those Si-based for instance) which are more mature and produced in an industrial environment according to well established procedures (Ito et al., 2007), for DSC a standard fabrication procedure is yet to be defined. Nevertheless the following steps are widely reported in literature, they strongly represent a prerequisite list to define and set-up an eventual industrial DSC pilot production line:

a - Cleaning of transparent and conductive substrates. Is generally carried out sonicating the substrates successively in solvents like acetone and ethanol for few minutes. In the case of glass substrates, this step may be further accomplished by firing the substrates in furnace or oven, in order to burn out the organic compound and preserve the subsequent processes.

b - TiO_2 film deposition. To obtain a mesoporous nc-TiO_2 film, from few micron up to few tens of microns thick, various techniques are adopted. One of the more diffused consists in preparing a colloidal paste composed of TiO_2 nanoparticles, organic binders, and solvents (Ito et al., 2007) and deposit it by various printing techniques such as screen printing, slot dye coating, gravure coating, flexographic printing, doctor blade and spray casting.

According to the printing technique performed, the composition of the paste and its recipe may observe some slight variation. For instance, in the case of automatic screen printer, it is recommendable to use TiO_2 pastes containing printing oil such as terpineol in order to facilitate the deposition process and a solvent like ethanol to optimize the deposition process for doctor blade technique. Also utilized are techniques such as spin coating, sputtering and electro deposition (Chen & Mao, 2007). It is interesting to point out that with the use of a layout or a mask it is possible to deposit the colloidal nc-TiO_2 layer according to a given pattern or shape.

Different authors have also shown the possibility to fabricate Dye Solar Cells utilizing TiO_2 films made with ordered nanostructures such as nanotubes, nanowires or nanorods (Chen & Mao, 2007). In these cases colloidal pastes are not anymore considered and furthers techniques are utilized. It is important to mention amongst the others, chemical vapor deposition, physical vapor deposition, sonochemical method and microwave method (Chen & Mao, 2007).

c - TiO_2 sintering. After the colloidal TiO_2 paste is deposited, a thermal treatment more often indicated as annealing, firing or sintering is required. In fact it is possible, via heat, to get rid of solvents and organic binders contained into the paste and in the meaning time to promote an electromechanical bonding intra-particles and between particles and the underlying substrate. This step clearly has a huge bearing on the film morphology and porosity which in turn determines the cells performances. At the end of the sintering process, a trade off is required between the necessity to guarantee a good electromechanical bonding and the requirement to keep a large porosity maximizing the sensitized surface area. Conventionally an optimum, depending on the paste composition, is obtained applying the photo-anodes (before the dye adsorption) to an increasing ramp temperature in an oven, furnace or hotplate with a final ~ 30 min step at 450–500 °C (Mincuzzi et al., 2009).

Although this procedure guarantees the fabrication of DSCs with good performances (included cells having the highest efficiency ever reported of approximately 12% over small area (Nazeeruddin et al., 2005; Buscaino et al., 2007)) it shows nevertheless some drawbacks. Since the nc-TiO_2 is heated with the substrate the mentioned conventional procedure is unsuitable for plastic substrates. For similar reasons DSC would not be integrated on the same substrate with others optoelectronic devices which would be destroyed by the high temperature. Oven and furnace are energetically expensive increasing the payback energy of the whole fabrication process. Finally, at 450–500°C conductive glass substrates could bend irregularly, preventing the possibility to fabricate large area DSC devices.

Alternative procedures have been proposed with the aim to overcome some of the mentioned drawbacks. For instance there have been several attempts to produce the TiO_2 film via low-temperature sintering suitable for plastic substrates (100–150°C) by utilizing a binder free colloidal TiO_2 paste. However, the devices with low-temperature sintered films were found to exhibit lower efficiencies than those with high-temperature sintered films. Pichot et al. (Pichot et al., 2000) have fabricated a flexible TiO_2 electrode that was spin coated onto indium–tin oxide (ITO)-coated PET substrates from an organic-free nc-TiO_2 colloidal suspension and then sintered at low temperature (100 °C) for 24 h. However, the overall device efficiency was relatively low (1.22%) under 1-sun illumination (100mW/cm^2). A mechanical compression of a surfactant free colloidal TiO_2 paste onto an ITO/PET substrate at room temperature has been demonstrated as an alternative sintering method for making plastic-based DSCs at temperatures between 25 °C and 120 °C. Utilizing this method, conversion efficiencies approaching 3% under 100 mW/cm^2 were reported (Lindstrom et al., 2002). Zhang et al. have developed a low-temperature chemical method for the fabrication of mesoporous TiO$_2$ films grown on ITO/PET substrates at 100 °C using a hydrothermal synthesis with conversion efficiency of 2.3% under 1-sun illumination. However, this process is only capable of partial crystallization of the starting Ti salts, and thus might require a subsequent high-temperature (450 °C) heat treatment (D. Zhang et al., 2003). Also several optical methods have been proposed including the use of I.R. lamps (Pan et al., 2009; H. Kim et al., 2006; Watson et al., 2010; Uchida et al., 2004) treatments. Nevertheless all the mentioned attempts produced solar cells with limited efficiencies compared to cells where standard high temperature sintering is carried out.

d - Dyeing. The nc-TiO$_2$ sensitization by dye adsorption (or dyeing) is carried out following two different strategies. In a case, sintered films are soaked with the substrate in a solution of ethanol and dye for a time period varying from around 10h to around 1h, depending on the dye solution temperature which correspondingly could be fixed in the range from T ambient to maximum T=80°C (which is ethanol evaporation temperature). Successively, TiO$_2$ films and substrates are rinsed in ethanol.

In the other case, devices are first sealed by means of a glass frit (see the step 6 below) and then dyeing is carried out fluxing the ethanol/dye solution through the gap between the electrodes. This second procedure has been introduced by the group of Fraunhofer ISE (Hinsch et al., 2008).

e - Pt deposition and thermal treatment. A crucial step in devices fabrication is to obtain a catalyst layer showing an effective catalytic activity. The main catalyst layer is a thin layer of Pt (some tenths on nanometers thick). Although the Pt layer could be obtained by sputtering, more frequently it is generally attained after thermal processing or curing of a Pt based precursor paste or solution which has been previously deposited by screen printing, doctor blade, spin coating or other printing techniques over the counter-electrode substrate. A Pt-based catalytic precursor paste composition suitable for printing technique is obtained mixing an organic compound (e.g. terpineol), a binder or stabilizer (e.g. ethyl-cellulose) and a precursor (e.g. hexachloroplatinic acid H$_2$PtCl$_6$) (Khelashvili et al., 2006).

A less viscous alternative, suitable for brush processing or spin coating, consists of an hexachloroplatinic acid solution in 2-propanol (Gutierrez-Tauste et al., 2005).

The cure of the catalytic precursor layer is conventionally carried out utilizing an oven, furnace or hot plate in order to treat the layer to a curing temperature with a final step of 5 – 30 minutes at 400 – 500°C (Kroon et al., 2007; Solaronix®). Whilst an effective Pt-precursor curing is guaranteed by this conventional procedure, nonetheless, the required high processing temperature induces some drawbacks already mentioned and listed in the case of nc-TiO$_2$ sintering. Kim et al. (S. S. Kim et al., 2006) also reported on Pt counter-electrodes prepared by means of direct current and pulse current electro-deposition methods, which are relatively simple and low cost processes.

Moreover these techniques are hardly scalable on large area devices. Alternatively, other materials than Pt, processed at low temperature (<150 °C) could be considered. As reported by Murakami et al., use of Carbon Black could be a valid and low cost alternative, and a small area cell showing power conversion efficiency exceeding 9% has been reported (Murakami & Graetzel, 2008). The same authors reported also a cell with catalyst layer made of Polymeric material such as PEDOT. Saito et al. and Suzuki et al. reported the application of a chemically produced conducting polymer and carbon nanotubes used as counter-electrode (Saito et al., 2002; Suzuki et al., 2003; Saito et al., 2004). Although the fabrication cost can be decreased in these cases, the conversion efficiency is still relatively low compared to a cell with a Pt based counter-electrode.

f - Device sealing. Ideally, device sealing should be long lasting, minimize as much as possible H_2O infiltration (detrimental for the dye), O_2 infiltration (it induces dye photo-catalysis) and electrolyte leakage, the principal cause of module contacts corrosion (Okada & Tanabe, 2006). The sealant has to be Pb free because Pb degassing results to be an hurdle for the Pt catalytic activity (Sastrawan et al., 2006). Find a sealing material and procedure fully satisfying all the mentioned requirement is still an open issue for the DSC technology.

Nevertheless, as already mentioned, two different strategies are generally adopted for this purpose, by using thermoplastic gaskets, or glass frits. In the first case the sealing is attained by thermal pressing the two electrodes with the gasket between them. Alternatively a lead-free glass frit paste is deposited between the two electrodes and then heated until it melts sealing the device (Sastrawan et al., 2006). The melting temperature varies upon the glass frit paste composition, although it generally exceed 400 °C. Whilst in the first case the process can be carried out having the photo-electrode already dye sensitized, this is prevented in the last case since the high temperature necessary for the glass frit melting will damage the dye molecules. As reported in the step 4, for this reason dyeing is in this case carried out by fluxing the dye solution (Hinsch et al., 2008).

g - Electrolyte injection. Finally, by back vacuum technique, electrolyte is injected into the sealed device through an hole drilled into one of the substrates. The hole is on end stopped by means of a glass patch. Further possibilities consist in fluxing the electrolyte (Hinsch et al., 2008) or in printing (before the device sealing) a previously gelificated redox mediator (P. Wang et al., 2003).

2.3 Large area devices

Large area dye solar cell devices are obtained interconnecting unit cells to form modules which could in turn be interconnected to realize a panel. The individual cells must not only be insulated from each other electrically (and this is performed via TCO laser scribing), but also electrolytically otherwise photo-induced electrophoresis would occur.

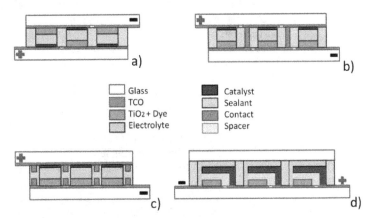

Fig. 6. Different architectures of Dye Solar Cells. Four different strategies are sketched: w-series interconnection (a), z-series interconnection (b), parallel interconnection (c), and monolithic-series design (d).

Furthermore, the electrical interconnections must be protected against the highly corrosive activity performed by electrolyte. As shown in Fig. 6 various interconnection architectures have been proposed for the modules, namely parallel, monolithic, z-series and w-series. Examples of modules obtained following different connections strategies are discussed in the literature (Tulloch, 2004) and disclosed in the following patents (Kay, 1995; Okada & Tanabe, 2006; Tulloch & Skryabin, 2006; Arakawa et al., 2007).

It is considered in particular modules to increase the voltage output cells which are connected together via either z-series or w-series schemes. In z-series design unit cells are sealed and connected by means of conductive vertical interconnections. The advantages of this design are an high voltage output and facility for eventual pre- and post-treatment of the electrodes. The disadvantage, is the risk of a fill factor lowering resulting from the series resistance of the interconnection. A crucial aspect regards the fabrication and realization of thermally stable vertical interconnections. Such interconnections need also to be protected from the corrosion eventually caused by the red-ox electrolyte leakage, which could compromise the modules performance and its working life time. An efficient and reliable interconnection strategy for z-series architecture modules is a still open technological challenge while some solutions have been proposed and disclosed in the following patents (Tulloch & Skryabin, 2006; Ezaki et al., 2006).

Differently from the z-series the w scheme avoids to interconnect altogether by juxtaposing cells facing in one direction with cells facing the opposite direction i.e. that have photo-electrode/counter-electrodes in opposite alternation and still requires separation of the cells by an effective sealing. This interconnection designs, is particularly attractive because the modules are potentially scalable to large dimensions avoiding the successive step of interconnection and integration of separate cells into a panel drastically simplifying the fabrication process compared to crystalline silicon panels. In DSCs the cells and interconnects can be integrated together by simple printing processes. Furthermore w-series design has advantages in simplicity and avoids the reduction in fill factor resulting from additional resistance of series interconnects, especially when the modules operate at high temperatures, but has some manufacturing and performance weaknesses. In manufacturing of this design, it is necessary that the counter-electrode and photo-electrode have to be processed on the same substrate. When conventional fabrication methods are utilized, this introduces processing complexities in deposition, curing, pre and post curing treatments of the cells materials in particular for the TiO_2 and catalyst precursor paste.

The design of a monolithic DSC (Vesce et al., 2011) is a planar-connected device with an only conductive glass, formed by a triple-layer sintered at once (Meyer et al., 2009) or together (Pettersson et al., 2007). The multilayer structure is typically composed by: a mesoporous TiO_2 working layer, a spacer/scattering layer and a counter-electrode layer. The spacer layer is often formed by ZrO_2 or rutile-TiO_2, while the counter-electrode layer is usually composed by Pt, Au, carbon or a mixture of these (Murakami & Graetzel, 2008). In order to obtain the maximum advantages from the monolithic architecture, some researcher reports the application of a quasi-solid gel electrolyte (non-volatile electrolyte) deposited by printing methods like blade-coating or screen-printing technique instead of an electrolyte processing by liquid injection (volatile electrolyte) (P. Wang et al., 2003).

For all types of serial DSC architectures, single cells must be insulated and serial connected causing a complex fabrication process, high serial resistance and a strong precision in order to maintain the ohmic contact (L. Wang et al., 2010). In contrast, a simple fabrication process is used to fabricate large scale parallel design DSC modules. In this architecture, (Tulloch,

2004), the grids utilizing conductive fingers to collect current are printed on the two substrates separately. Nowadays, the researchers that work on this design are devoting to find alternative materials solutions for the grid since Ag, Au, Cu, Al, and Ni are all easy to be corroded by the iodide electrolyte (Goldstein et al., 2010). Moreover, new coatings like polymer or glass frits are successfully applied in parallel DSC module design.

Having in mind the materials, strategies and fabrication processes aforementioned, various research groups (CHOSE, ISE Fraunhofer, KIST, NREL) and companies (3GSolar, G24i, Sony, Fujikura) have shown DSC panels and large area devices utilizing fabrication process which could be potentially integrated into an automatic production line.

Nevertheless, the scale up of DSC devices over large area for industrial and commercial purpose, poses some scientific and technological challenges. In particular, it is important to mention the reduction of the series resistance (Dai et al., 2005), strictly related to the increasing of the performance, stability and working life time of large area modules (P. Wang et al., 2003; Sommeling et al., 2004; Kuang et al., 2007), and the setting-up of a highly automated and energetically efficient fabrication process (Somani et al., 2005; Meyer et al., 2007; Mincuzzi et al., 2009).

3. Characterization techniques

The typical parameters to characterize a PV device belong also to DSCs performance. However, different technologies often imply different attentions to perform the measurement. Parameters like short circuit current, open circuit voltage, fill factor and power conversion efficiency, are extracted from the photo-current/voltage characteristic (J-V curve). This measurement can be fundamentally performed in two ways: in dark conditions (without illumination) and with illumination. In the last case, we are able to extract all the parameters. Illuminating with white light, the J-V curve represents an integrate response of the cell to all the wavelengths simultaneously. The standard test conditions (STC) correspond to power density of 1 sun (100mW/cm^2), light spectrum of an air mass 1.5 global (AM1.5G) and working temperature of 25°C.

Incident Photon-to-electron Conversion Efficiency (IPCE) measurements, on the other hand, quantify how many incident photons at a single wavelength are converted in extracted electrons at the electrodes. It strictly depends on the optical and electrical properties of the cell. In particular, it can be decomposed in the product of three main terms: light harvesting efficiency (LHE), injection efficiency (η_{inj}) and collection efficiency (η_{col}) (Halme et al., 2008; Barnes et al., 2008). The first is reduced by optical phenomena, such as reflections, transmission and competitive absorption, while injection and collection efficiencies depend on electrical mechanisms at widely different time scales. The injection counts how many excited electrons are transferred from dye molecules to titanium dioxide film. It is the fastest process in the cell, with characteristic times from femto to pico-seconds, measured by means of pump-probe techniques (Koops & Durrant, 2008).

The collection efficiency says how many injected electrons into the titanium dioxide film reach the electrodes. It can be evaluated by many different techniques, among which photovoltage decay (Walker et al., 2006), intensity modulated photovoltage spectroscopy (IMVS), intensity modulated photocurrent spectroscopy (IMPS) (Schlichthorl et al., 1999) and electrochemical impedance spectroscopy (EIS) (Q. Wang et al., 2005). The slowest process in the cell is the ionic diffusion in the electrolyte, which can be studied by EIS as well.

3.1 I-V characteristics and IPCE spectra

I-V characteristics and IPCE spectra can be related each other. Actually, from the complete IPCE spectra, it is possible to calculate the short circuit current that one should measure under solar simulator in the following way:

$$J_{SC}(\lambda) = \int_{\lambda 1}^{\lambda 2} IPCE(\lambda) P_{STC}(\lambda) \frac{\lambda}{1240} d\lambda \qquad (1)$$

with $P_{STC}(\lambda)$ the spectral density at standard test conditions and λ the wavelength in nm. This calculation is an important check that should be always performed to control the measurement accuracy. Many factors can influence the deviation of the current measured under solar simulator from the value calculated by equation (1). In particular, according to the solar simulator class, the spectrum can considerably differ from AM1.5 conditions (Ito et al., 2004). Because the calculation is performed at AM1.5, an overestimation or an underestimation can occur. It mainly depends on the light harvesting efficiency of the cell or more simply from the absorption coefficient of dye molecules. A good way to overcome this discrepancy consists in estimating the mismatch factor (M) (Seaman, 1982). It considers the spectral responses of the test and reference cells and the spectral irradiance of the solar simulator (while AM1.5 is known). It can vary a lot among various kind of cells (it is close to unity for silicon solar cells). In Fig.1 the calculated and the measured currents for dye solar cells with different dyes are put in relation. Different dyes mean different absorption spectra, and consequently a variation of the mismatch factor. Although a linear relation is obtained, the angular coefficient is different from one. Moreover, to make the things more complicated, there is the dependence of dye solar cells response from the level of illumination. So, as it will be clear soon, to relate the results from IPCE and J-V measurements is important to be in the same light intensity conditions. It means that also IPCE spectra should be acquired at 1 sun condition.

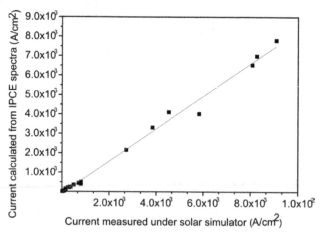

Fig. 7. Density Current measured under solar simulator and calculated by IPCE spectra are put in relation for cells with different dye molecules. A perfect agreement would give an angular coefficient m equal to one. In this case we have m=(0.83±0.02) and an adjusted R-square of 0.992.

Actually, DSC photovoltaic characterization is critical. Performing J-V curve, the direction of scan as well the delay time during the measurement must be chosen accurately otherwise different results can be obtained. One of the most important reason for these different behaviors is due to strong capacitance effects presented in this kind of device (Koide & Han, 2004). The main consequence is the long constant time of this kind of cells (in the order of some seconds) with respect to other technologies. An overestimation of short circuit current can be carried out, in particular when small area cells are characterized. In this case, the device area is generally larger than the active area, and, when illuminated, a considerable amount of light not impinging onto the active area can be redirected to it (light piping effect) (Ito et al., 2006). According to the simulator class, the beam divergence can amplify this effect. To overcome it, an appropriate opaque mask must be applied onto the external surface front glass. Then, particularly for large area devices, or for devices delivering high current, the external bad contacts can strongly influence the measurement. Good contacts can be obtained with bus bars applied by screen-printing technique.

On the other hand, IPCE measurement on dye solar cells is a critical issue as well. IPCE measurements can be performed in two ways, applying a direct (DC) or an alternate (AC) method. The first one is the classical way to acquire IPCE spectra, while the second one consists in illuminating the cell with white light (also called bias light) simultaneously with the monochromatic component. The bias light acts as a sort of polarization of the cell, increasing its response, besides the fact that, in this way, the cell can be put under conditions closer to the working ones. The current due only to monochromatic light (we say monochromatic current) is discriminated from the current due to the bias light, by using a coherent detection. It means that the monochromatic light is modulated at a certain frequency and by a lock-in amplifier, only the current modulated at the same frequency will be detected.

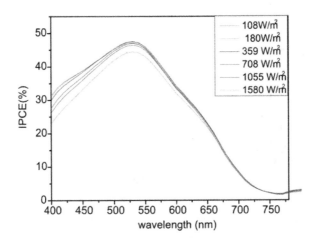

Fig. 8. IPCE spectra in function of the bias light illumination. A clear dependence from the light power density is shown. In the legend, the bias light power density is shown.

There are mainly two effects affecting IPCE when we illuminate with different power density conditions: the trap filling effect and the electrolyte ions mobility. While the first affects negatively the IPCE spectra at low light level conditions, the second comes into play at high light density reducing the solar cell response as well. For trap filling we mean the ability to occupy the states inside the titanium dioxide gap, close to the conduction band edge. These levels are centers of recombination for the electron in conduction band. At single wavelength, the filling is not efficient, reducing the cell response (see Fig. 8). It has been verified that the application of a bias light can be simulated in the DC method, if the intensity of the monochromatic light is high (Sommeling et al., 2000). On the other hand, at high intensity the electrolyte ions could be not able to regenerate effectively the homo level of the dye. This effect is dramatically enhanced when we use $Co^{(II)}$-$Co^{(III)}$ as redox couple.

Aware of the dependence from light intensity, to control the measurement accuracy under solar simulator, it is mandatory to perform IPCE acquisition at the same conditions.

Different dynamics are present in the photovoltaic mechanism of a dye solar cell. In presence of illumination, however, only the slowest process will dominate. The result is that the dye solar cell response is really slow. The modulation of the monochromatic light should be less than 1 Hz, taking into account that it should be verified every time different materials are involved (in particular the electrolyte and the titanium dioxide film employed).

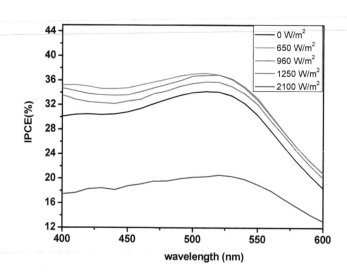

Fig. 9. IPCE spectra in function of the bias light illumination for a dye solar cell with $Co^{(II)}$-$Co^{(III)}$ as redox couple. A decrease of the signal intensity at high intensity levels has been measured.

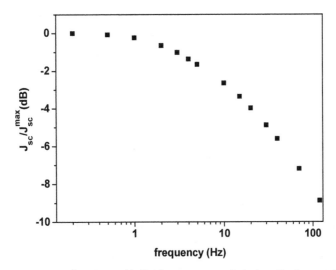

Fig. 10. DSC response in function of light frequency modulation. In the y-axis, photocurrent appears as a ratio (in decibel) with respect to the maximum current at the lower frequency. The response of a DSC is slow if compared to silicon technology.

IPCE spectra take in account many different phenomena that we can distinguish in two main categories: optical and electrical ones. In particular IPCE depends on the ability of the cell to harvest the light. Photon management techniques try to improve just this factor. The light harvesting efficiency of the cell can be calculated starting from spectrophotometric measurements. A simple optical model of the geometry allows the estimation of this quantity, that is the electrons generated compared to the incident photons. In a simplified scheme, assuming a Lambert-Beer behavior, we can model the light harvesting efficiency when the light impinges onto the front side of the cell in the following way:

$$LHE(\lambda) = T_{TCO}(\lambda) \cdot \frac{\alpha_{dye}}{\alpha} \cdot \left(1 - e^{-\alpha d}\right) \qquad (2)$$

where T_{TCO} is the transmittance of the transparent conductive oxide, α is the absorption coefficient of the entire film and α_{dye} is the absorption coefficient due to the dye molecules. This is, obviously, a simple approach, where second-order reflectance terms are not considered. Measuring IPCE and estimating LHE, we are actually able to obtain information about injection and collection efficiencies just making the following ratio:

$$APCE(\lambda) = \eta_{inj}\eta_{col} = \frac{IPCE(\lambda)}{LHE(\lambda)} \qquad (3)$$

where APCE stands for Absorbed Photon to Current conversion Efficiency and it is the product between injection and collection efficiencies.
Making the measurements illuminating both sides of the cells in different times, an estimation of the collection efficiency, the diffusion length (L_D) and the injection efficiency, has been demonstrated under strict conditions (Halme et al., 2008; Barnes et al., 2008).

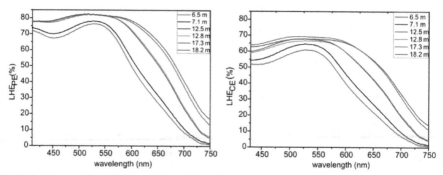

Fig. 11. Light Harvesting Efficiency for cells with different thicknesses illuminating from photo- (on the left) and counter- electrode (on the right) sides.

In Fig. 11, estimation of LHE for different thicknesses of the titanium dioxide film for both directions of illumination has been reported. As intuitive, LHE from counter electrode side is typically less than in the case of front side because of the generation profile inside the titania layer and the electrolyte absorption, mostly in the wavelength range under 500 nm.

4. Photon management

The typical paths followed to increase the performances of DSCs are linked to their main components, i.e., to improve the mesoporous nanocrystalline titania (nc-TiO$_2$), to find new dyes or dye combinations and to improve the ionic electrolyte. Approaches to enhance efficiency are also being followed which belong to a wide strategy of photon management. The dye management itself acting on the dye properties may be considered inside the panorama of photon management (Park, 2010). It consists in a multiple dyes co-sensitization in order to enlarge photonic response via panchromatic absorption, hence to increase efficiency. There have been already proposed works focalizing on the panchromatic feature of a dye solar cell (Ogura et al., 2009; Yum et al., 2007; Park, 2010). The way to get improvement is by the use of two (up to three) dyes adsorbed on the nanocrystalline titania that are responsible for broad spectral response of the device. The development of organic sensitizers (C101 etc.) (C.-Y. Chen et al., 2007; Abbotto et al., 2008) led to very high levels of efficiency. More in general, photon management consists in the ability to confine light in the dye solar cell to stimulate high levels of charge enforced by scattering and reflection effects. At the same time, this should be coupled to decreasing the recombination of charge mostly at the interface nanocrystalline TiO$_2$/electrolyte. Indeed, it is known that the top performances of DSC devices are reached by keeping in mind also all the parasitic and recombination effect and the way to minimize them. For example, in order to quench the recombination at FTO/electrolyte interface and to facilitate the injection between the dye LUMO and the TiO$_2$ conduction band, it can be used a photoanodes treatment by a titanium tetrachloride (TiCl4) solution (Vesce et al., 2010). Then, the transparent layer of titania (average particle diameter 15-20 nm) can be covered or added by larger scattering particles (150-400 nm in size) (Usami, 1997; Arakawa et al., 2006; Colonna et al., 2010) causing the random reflection of the light back into the cell (Mie scattering). Indeed, the most common way of photon management consists in the development of diffuse scattering layers (SLs) capable to be used as incoherent back mirrors for the incoming light passing through the cell and otherwise not converted into current. In 1997 (Usami, 1997) a theoretical work by A. Usami proposed the use of a scattering layer onto the nc-TiO$_2$ layer and a rutile thin layer between the glass and TCO conductive film. This implies a very effective enhancement of the light collected into the cell, but also means that the DSC remains opaque. Nowadays, the scattering layers (Hore et al., 2006; Arakawa et al., 2006), centers (Hore et al., 2005) and superstructures (Chen et al., 2009; Q. F. Zhang et al., 2008) are well known and routinely used (Graetzel, 2005). Despite other approaches to the problem of increasing DSC performances while maintaining light transmittance (Colodrero et al., 2009a; Ogura et al., 2009) the record of performance for a DSC is obtained by the use of diffuse SLs (Nazeeruddin et al., 2005; Arakawa et al., 2006). To confer order to the scattered light, Miguez proposed the selective mirror for DSC (Colodrero et al., 2009a) made out from colloidal TiO$_2$ suspensions (Wijnhoven & Vos, 1998; Colodrero et al., 2008). They consist in photonic crystals (PCs) (Yip et al., 2008; Colodrero et al., 2009b), introduced either inside the titania layer or on its backside (Nishimura et al., 2003; Mihi et al., 2006), currently under an intense experimentation. Scheme in Fig. 12 resumes some of the light management approaches for conversion efficiency improvement.

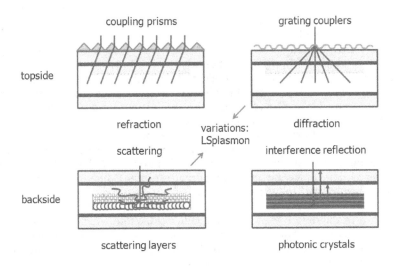

Fig. 12. Photon management basic approaches.

Some of these techniques will be described in the following sub-sections. In both SLs and PCs techniques of photon management, the increased light path in the active layer (e.g., by scattering or interferential confinement), will enhance the light harvesting efficiency (LHE). Even the reflection can be exploited to call into play of photons otherwise lost from the cell, as in V-shaped or folded solar cells (Tvingstedt et al., 2008; Zhou et al., 2008). In the waveguide DSC (Ruhle et al., 2008) a coupling prism let the light enter beyond the condition of total reflection at the glass plates/air interface without letting it to escape. Plasmonic solar cells (Tvingstedt et al., 2007; Catchpole & Polman, 2008) may represent another kind of photon management for field enhancement (near-field) or scattering by surface plasmon polaritons (mostly localized on metallic nanoparticles). Other configurations involve field enhancement plus diffraction from metallic subwavelength arrays (Hagglund et al., 2008; Pala et al., 2009; Ding et al., 2011). An increased optical path may be obtained in principle also by dielectric diffraction or refraction (Dominici et al., 2010). Structuring the top side with a dielectric surface texturing, either nanometric or micrometric (Tvingstedt et al., 2008), could achieve the additional (diffracted) light rays or a larger inclination of (refracted) path (respectively by using of grating couplers or microprisms and microspheres for example).

4.1 Co-sensitization

The co-sensitization of nc-titania anodes approach consists in the use of two or more dyes anchored on the same substrate (Chen et al., 2005; Shah et al., 1999). It has been considered with particular attention to some organic dyes having complementary spectral response in the red with respect to the ruthenium-based dyes (largely used for standard DSC), such as squaraine (SQ1) (Clifford et al., 2004), cyanine (Pandey et al., 2010), phthalocyanine (Ono et al., 2009), hemicyanine (Cid et al., 2007). Indeed in other studies the co-sensibilization approach has shown high device performances toward red and violet as well in the electromagnetic spectrum (Yao et al., 2003; Kuang et al., 2007; Yum et al., 2007, 2008; Chen et al., 2005; Clifford et al., 2004). The scope of co-sensitization is to enlarge the absorbance spectrum of the cell toward the Near Infra Red (NIR), thus to increase the Incident Photon to Current Efficiency (IPCE) by enhancing the LHE (*Light Harvesting Efficiency*) and the efficiency of injection inside the TiO_2 (see IPCE section).

Here have been investigated the co-sensitization effects by using two conventional Ru-based dyes, the N719 and the Z907, together with a second one that is a typical Dye for dye lasers (HWSands). With respect to other co-sensitization approaches it has been shown the improvement of performances without losses when the dyes are both anchored to TiO_2. This means that the behavior of photocurrent and efficiency is summed not linearly, i.e. more than the sum of each single dye performance cells.

The most important fact to take into account in this approach is that the dye does not reach the saturation point, i.e. maximum allowed absorbance and hence maximum performances. What done is the immersion by using the first ruthenium dye followed by the second one for a determined time. In fact by setting properly the dipping time there have get enhanced performances with respect to 'one dye system DSC'. It should be noted that the immersion time far from the saturation of the titania layer for the ruthenium dyes implies technological reasons. In fact in Building Integrated Photovoltaic (BIPV), to which DSC are devoted, the transparency is a central factor. A saturated working electrode will be slightly opaque, while by using a second dye absorbing toward the red together with the unsaturated one is possible to keep an acceptable level of transparency and efficiency.

Experimental spectra were acquired with the integrating sphere of a Spectrophotometer by using the undyed titanium dioxide substrate as reference. The working electrode's absorbance saturates after some hours for N719 and Z907 depending on the thickness of TiO_2 and dye concentration whereas for SDA is found that the saturation time is of the order of 15-30 minutes for both thicknesses investigated and has been also observed a photo-cleavage due to TiO_2. In the figure below are reported absorbance of N719 on nc-TiO_2 at different times and the photocatalisys of NIR dye.

The optical response of the double dye is enlarged up to 700nm due to the presence of near IR dye. It should be noted that prolonged dipping time in the SDA solution will cause a displacement towards the N719 molecules already attached on the TiO_2 surface; in fact MLCT (Metal to Ligand Charge Transfer) band absorption of N719 (3h) decreases after 15 minutes dipping in SDA. The same trend is kept also for 30 and 45 minutes (see Fig. 14).

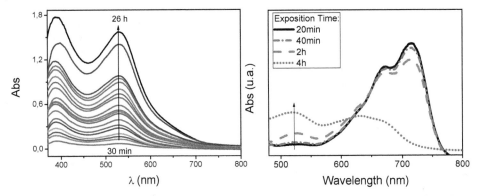

Fig. 13. (Left) Absorbance of nc-titania dyed with N719 (30 min up to 26 hours) and (right) photo-cleavage of SDA due to the TiO_2.

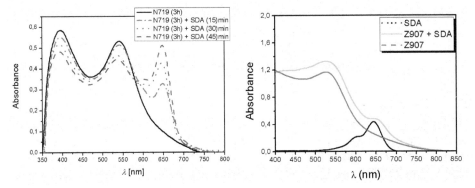

Fig. 14. Left: Co-sensitized spectra of the SDA1570 dye together with N719 on nanocrystalline titania substrates (6 μm) along with single dye absorbance. Several dipping times were chosen to show the decreasing peak of the N719 due to SDA1570 effect. Right: Co-sensitized spectra of the SDA1570 dye together with Z907 on nanocrystalline titania substrates (12 μm) along with single dye absorbance.

There is the gradual detaching of the N719 molecules from the titania due to the SDA environment. In this process it should be considered the equilibrium constants of the process involving initially the N719-TiO_2 photoelectrode in EtOH solution of SDA. The latter molecules act on the substrate by mass action due to the concentration gradient. The SDA molecule acts for N719 detaching from the TiO_2 surface. This depends mainly on the concentration of SDA solution, on the temperature, and the time. Finally there will be reached a dynamical equilibrium in which the number of SDA entering molecules on titania is equal to the same detaching molecules. Since such configuration is undesired, the finding of the optimal adsorbing point by both N719 and SDA molecules is central factor.

For completeness the action of SDA on dyed N719 PEs and *vice versa*, immersed up to 18 hours on titania was investigated (see figure 15, right). It is found that SDA is not able to detach all the N719 molecules, consequently the absorbance has almost the same trend for

15 minutes and 18 hours of SDA on saturated (18 h) N719 PE. The N719 instead shows an increasing of the absorbance passing from 15 minutes to 18 hours when alone (figure 15, left); moreover the attachment dynamic of N719 is very slow if compared to SDA. On the contrary it can be seen that the N719 environment for a saturated SDA photoelectrode is deleterious for the latter, being completely cancelled (figure 15, dot curve). It can be noted that the maximum absorbance of N719-SDA PEs is almost the same for 15 minutes and 18 hours of SDA immersion meaning that the affinity of SDA to the N719 saturated titania is limited.

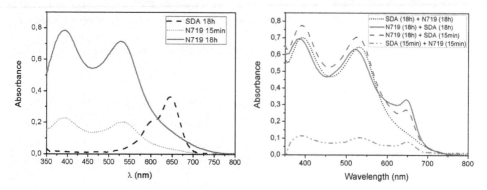

Fig. 15. Absorbance of 6 micrometers titania PEs in several dye adsorption configurations; (left) single dye TiO₂ attachment and (right) saturation conditions.

A similar study for Z907 + SDA system has been carried out; the transparent 12 micrometers thick TiO₂ PE was dipped in Z907 (0.3 mM) for 5 hours, while SDA for 30 minute steps. In this case, due to the ability of the thicker PE to generate an higher current with respect to the previous case, the electric performances are notably higher than N719 (Fig. 16).

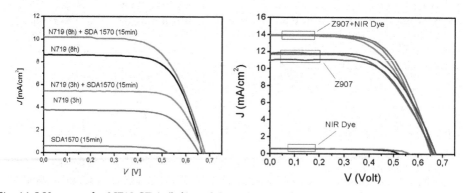

Fig. 16. J-V curves for N719-SDA (left) and Z907-SDA (right) co-sensitized systems. The lowest curve is due to the SDA sensitizer alone (labeled NIR in the right plot). It can be seen that the contribution of SDA is very small when compared to the N719 or Z907 current generation, but it becomes very important when the ruthenium dye is already and partially attached to the surface.

In this case, by taking into account that the Z907 Ruthenium-based dye has hydrophobic chains, we shall consider that (relatively) prolonged dipping times are required by the SDA to attach efficiently to the Z907 dyed titania PEs. This explains the small absorbance seen in figure 1 where the Z907 (5h) is immersed for thirty minutes in SDA solution.

The cells assembled by using the above photoanodes arrangements have been tested under a sun simulator (AM1.5) at 0.1Wcm^{-2} of illumination density of power. It is found that for N719-SDA system (at different dipping times) the co-sensitized cell outperform the single dye, having unexpected Jsc generation and efficiency. The same trend, but with higher values, has been found for Z907-SDA arrangement.

The Internal Photon to Current Conversion Efficiency confirms the above trends showing a zone of generation at the SDA excitation energy (650-660 nm).

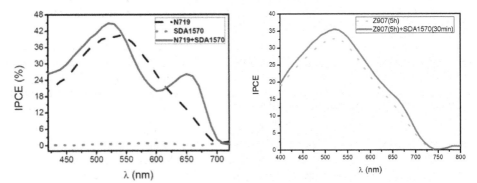

Fig. 17. IPCE results of the studied systems. In the case of N719-SDA couple the SDA pick is well identified at 660, whereas in the Z907-SDA only a small increasing of the IPCE figure is registered.

The immersion of the partially N719-sensitized photoanode in a SDA solution induces the saturation of the remaining free TiO$_2$ surface and at the same time a partial displacement of the already attached N719/Z907 molecules, creating a sort of "self-organization" of the two molecules that improves the cell performance, limiting the energy loss due to excitonic interaction between homologue molecules. This seems to be confirmed by IPCE measured. It shows in fact that photocurrent for the co-sensitized cell has a relative maximum in the wavelength region of maximum absorbance of SDA1570 confirming that it acts as an absorber on the TiO$_2$ but not as carrier generator in the cell when anchored alone to the titania. Instead, if attached together with N719 a major contribution in charge collection starts. Moreover the N719 active spectra in the co-sensitized device is blue shifted and narrower than that in the non co-sensitized device. Such a molecular organization effect can justify the fact that SDA1570 alone is not a sensitizer, while together with N719 it becomes a sensitizer for DSCs (Colonna et al., 2011a).

4.2 Diffusive scattering layers

The use of larger titania particles dispersed or added in layers on the nc-TiO$_2$ slab of a dye solar cell has been proven to be the best arrangement for high performance DSC (Nazeeruddin et al., 2005). The scheme of a DSC having a thin slab of opaque titania particles (~ 150-400 nm) onto the transparent one in several configuration is depicted in Fig. 18. The optimal diameter of the transparent nc-titania particles is about 15-20 nm; during the sintering process at nearly 500°C, the particles create the mesoscopic structure and the effective surface of the TiO_2 electrode is increased by up to 10^3 factor with respect to the apparent area. In this way when the dye is adsorbed there are up to 1000 monolayers of dye in the cell for charge generation (Ferber & Luther, 1998). The pores in the layers have the better diameter for electrolyte infiltration and diffusion. If the TiO_2 particles are too small, the pores are not large enough for the dye and the electrolyte infiltration. Finally the larger the size particles the smaller is the internal surface, hence poor charge generation.

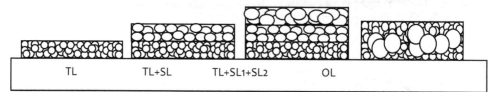

Fig. 18. From left to right hand: few micrometers nc-TiO_2 (~ 15-20nm); single scattering layer (d ~ 100nm) on the previous; double scattering layer with upper one having d > 200nm particle size; dispersion of small and large diameter TiO_2 particles. TL = Transparent layer, SL = Scattering Layer, OL = Opaque Layer.

Due to the opacity of scattering titania particles placed onto the transparent nc-TiO_2 the incident light passes through the nanocrystalline dyed titania, then it encounters the diffusive slab of bigger particles and is resent back to the PE finally. The average size of the scattering particles can be tailored to be between 60 and 500 nm, whereas the layer thickness can vary between 3-4 and 20 micrometers (Arakawa et al., 2006; Koo et al., 2008).

It should be considered that by doubling the thickness of nanocrystalline transparent titania the photocurrent will not be doubled because the difference in transmittance decreases with increasing wavelength, that is, little difference at wavelength ranging from 650 nm to 800 nm. For this reason, a TiO_2 film having only nanocrystalline particles cannot improve photocurrent density significantly by increasing the film thickness (Park, 2010). For this reason the random effect of a diffusive layer can enhance the reflectivity back to the cell by increasing the incident light path length and therefore the absorption, thus the LHE. All the works based on such strategy have been based on A. Usami (Usami, 1997) studies to demonstrate that with a simple model for multiple scattering the best configuration can be obtained with particles which size is a fraction of the incoming wavelength. Usami considered that Mie scattering theory is a rough approximation if scattering particles are not spherical and for multiple scattering. To take them into account some corrections have to be introduced. The exact solution of scattering of light by a particle is obtained by Mie theory, along with the dependence on particle size, absorption index, uniform dispersion of the particles, sufficient particle condensation for effective electron transfer and sufficient opening for the adsorption of the sensitizers (Arakawa et al., 2006; Park, 2010).

It has been found that the optimal scattering matching condition is obtained for $kd/\pi = 0.7 \sim 1.6$. Since the wave vector is given by $k = 2\pi/\lambda$, this condition implies that it exists an interval of wavelengths and size scattering particles for best improvement condition.

For this study it has been investigated firstly the absorption, i.e. $A = 1 - T - R$, of substrates taking into account the reflections of the device. In this way can be understood the spectral area in which the diffusive layers can efficiently operate. In the figure below can be seen the absorption of nc-TiO$_2$ of 6 and 12 μm along with the SLs effect. It should be noted that the growth of 1 or 2 diffusive slabs of the same particle diameter creates the same absorption to the PE.

For quantitative estimation on the cell performance the study the IPCE trend of the cell is required in order to see explicitly the enhancement factor. This is because the absorption curve does not take into consideration the final device arrangement, that is the current generated by itself. On the right plot of the figure are shown reflectance spectra (diffuse and specular) due to transparent or scattering particles, in a normal configuration. Typically the SL can enhance the photocurrent to very high percentage because of the random reflection. Indeed it can be seen that almost all the reflected light by the scattering layer is intercepted by the dye pigment up to 600 nm. Therefore the absorption A of the cell will be increased as the IPCE.

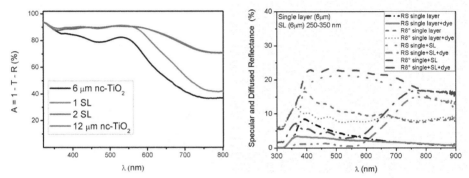

Fig. 19. (Left) Absorption of nc-TiO$_2$ dyed electrodes and the same covered by one or two diffusive scattering layers. (Right) Diffuse and specular reflectance of the 6 μm titania added by the dye (N719) and not.

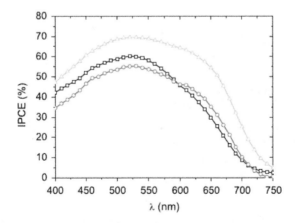

Fig. 20. External Quantum Efficiency of a standard DSC along with two scattering arrangements.

Tipically the IPCE curves have the shape reported in the following figure. In that case thicknesses for both transparent and opaque layers are reported in table 1 (Colonna et al., 2010). The enhancement in the zone over the dye pick has been simply obtained, confirming the idea followed from the above discussion (Usami, 1997).

The electrical values registered are shown in the table. The photocurrent reaches an increment > 45% by using a scattering layer with the same size of the transparent slab, whereas it is quenched by a thicker scattering layer (~ 22 μm).

Finally it is instructive to evaluate the enhancement factor due to the ratio between IPCE$_{SL}$ and IPCE$_{St-DSC}$. The region of the actual enhancement due to the scattering layer is centered at over 700 nm.

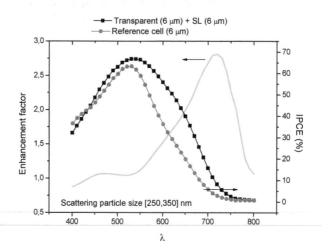

Fig. 21. Single IPCEs for transparent and SL arrangement and enhancement factor due to the SL.

Cell	TiO$_2$ (μm)	Jsc (mA/cm^2)	Voc (mV)	ff (%)	η (%)
A	12	13.7	710	61	5.93
B	12 + 12	20.2	719	59	8.55
C	12 + 12 + 10	14.4	766	59	6.53

Table 1. Arrangement of photoelectrodes and electric performances of standard DSC (st-DSC) along with scattering structure integrated.

In conclusion the use of larger titania particles for light scattering within the dye sensitized solar cell has been investigated in terms of enhancement in the red region of the spectrum. It has been found that for particular arrangements the photocurrent improvement can reach unprecedented results (Colonna et al., 2010).

4.3 Photonic crystals

One Dimension Photonic Crystals (1DPC) within the DSC assembly represent probably the most important field for future development of the field for several reasons soon described. Up until 2008 it was known from some authors that the integration with inverse opals

(3DPC) could be possible but mechanisms arising in that dye solar cell is different from the one described in the rest of the section since it is coherent scattering effect (Nishimura et al., 2003; Halaoui et al., 2005; Lee et al., 2008; Mihi & Miguez, 2005; Mihi et al., 2006, 2008). The combination of one dimension photonic crystal (1DPCs) layers made by using colloidal solution of SiO_2 and TiO_2 in the dye solar cell technology has been introduced by S. *Colodrero* and *H. Miguez* at CSIC in 2008 as a new powerful tool for DSC technology (Colodrero et al., 2008). They demonstrated the physical properties of the photonic crystal stack in terms of modes of the light once has passed through the multilayer assembly (Colodrero et al., 2009a; Colodrero et al., 2009b; Lozano et al., 2010). The materials integrated on the nc-TiO_2 is composed by alternating SiO_2 (n_{SiO2} ~ 1,5) and TiO_2 (n_{TiO2} ~ 2,5). The periodic arrangement of layers creates patterns of waves interfering in a range of wavelength depending on the thicknesses of each layer when the light is reflected. This imply that the DSC-PCs based can generate a gain with respect to a standard DSC because both the incoming polychromatic light stimulates transitions (standard process) and the reflected PC's band is sent back into the cell. Moreover the arrangement of silica-titania bi-layers creates a periodic refractive index responsible of the photonic band, causing the structural color of the photoanode (Calvo et al., 2008). The Bragg's law implies that the reflected wavelength due to an *optical thickness* of $n_1d_1 + n_2d_2$ is:

$$\lambda_B = 2(n_1d_1 + n_2d_2) \tag{4}$$

where the factor 2 refers to the double verse (in and out) of the optical path. The emission photonic band can be calculated by considering the Distributed Bragg Reflector (DBR) used for waveguides. In this case the *photonic stop band* is given by the formula:

$$\Delta\lambda_B = \frac{4\lambda_B}{\pi} asin\left[\frac{n_2 - n_1}{n_1 + n_2}\right] \tag{5}$$

where λ_B derives from the (1). This band represents the optical range of reflected wavelength on the alloy and for the materials used in this study for example with a λ_B = 650 nm, the $\Delta\lambda_B$ is ca. 200 nm (see figure 2). The intensity of the reflectance is given by:

$$R = \left[\frac{(n_0)(n_2)^{2N} - (n_S)(n_1)^{2N}}{(n_0)(n_2)^{2N} + (n_S)(n_1)^{2N}}\right]^2 \tag{6}$$

where n_0, is the refractive indexes of the entering medium, n_1, n_2 the indexes of the alternating materials and finally n_s the index of the exit material. N is the number of the bi-layers creating the structure.

The PC can be created with a simple reliable procedure (Colodrero et al., 2008) giving the possibility to tailor the optical thickness by varying the operative settings of deposition. It consists in the growth of layers by spin coating technique. The final result is the creation of an stack of porous layers. Due to the porosity itself the electrolyte can infiltrate in the pores where it modifies the dielectric constant, hence causing the variation of refractive index of the layer stack and the reflectance Bragg's peak is consequently red shifted according to the Eq. (4). Therefore the reflectance of the complete DSC device will present reflection at wavelengths corresponding to the previous reported in figure plus a shift to the red because of refractive index variation. The reflection will enhance electrical and optical characteristics of the standard cell by conferring selective photocurrent enhancement. Indeed the IPCE shows well defined improvement zones corresponding to the reflected range of light.

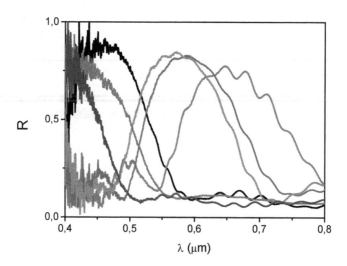

Fig. 22. Reflectance on nc-TiO$_2$ PEs containing SiO$_2$/TiO$_2$ bi-layers measured by FTIR.

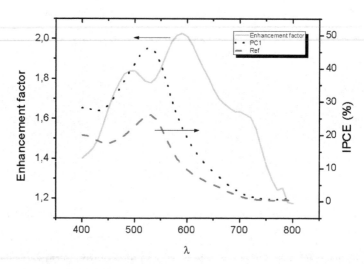

Fig. 23. IPCE enhancement factor calculated by the ratio between the PC integrated and the standard DSC.

The last point is of importance in this development because not only the cell will have high performances, but also a structural coloration will arise independently of the dye color. Finally the important consideration is that the 1DPC-DSC keeps the transparency, meaning that such DSC branch can be further explored for BIPV applications (Colonna et al., 2011b).

4.4 Angular refractive path

Recently (Colonna et al., 2010; Dominici et al., 2010) a strong enhancement of short circuit photocurrent I_{SC} by varying the angle of incidence of a monochromatic laser beam was shown for DSCs. A light path lengthening is active, supposedly, due to the typical features of the absorbing (titania) layer in the semitransparent DSC. I.e., its (relatively) low refractive index n and absorption coefficient a which offer margin improvement for an Angular Refractive Path (ARP) factor to increase the LHE. Indeed, an external oblique incidence θ_a of light corresponds to an oblique angle of propagation θ_{eff} inside the sensitized titania too. The lower the effective index n_{eff} the larger the internal angle θ_{eff}. When $a \cdot h$ is low, an inclination of the propagation line inside the active layer allows to lengthen the path and further absorb light beyond the inherent limit of the native thickness h. Evidence of the ARP factor depends both on the thickness of the cell and the wavelength, plus the eventual use of a coupling prism. The prism allows indeed to reach larger angles of propagation. According to theory, the ARP is shown to be more effective for thinner cells and at wavelengths where the dye molecules absorb less, while the use of the prism enhances it further. The ARP may also explain why DSCs under diffuse illumination work better than other PV technologies, giving hints for new concepts in design of more efficient DSCs.

In order to present evidence of such effect, we initially propose three simple configurations in Fig. 24. The same cell is firstly illuminated in an EQE (i.e., IPCE) setup at θ=00° (normal incidence) retrieving the quantum efficiency spectrum. Then the DSC is rotated and illuminated at a θ=45° angle of incidence. Hence, for the same angle in air (between the light beam and the normal to the cell) a coupling prism is used (half cube, BK7 glass prism). In this last case a matching index oil (n=1.66) is used between the prism (n=1.515) and the glass substrate (n=1.59).

Fig. 24. Three simple configurations to test the refractive angular path. They correspond to normal incidence (θ=00°) without prism, oblique incidence (θ=45°) without prism and oblique incidence (θ=45°) with prism. To keep the light spot always wholly inside the active area means to have constant impinging power. The external reflections are represented together with reflection from the active layer.

The spectra registered in the wavelength range 400-650nm appear in Fig. 25, from bottom to top following the order of their presentation. There is a certain enhancement deriving from the use of an oblique incidence, further pushed up by the use of the prism. Such enhancement can be represented by normalizing the last two curves to the first one. It presents two main features. Firstly, where the EQE (hence, absorption) is high the enhancement has got a local minimum and vice-versa. This feature is expected as introduced on the basis of the ARP theory, discussed more in detail in the following. Secondly, there is a certain monotonic increase of the enhancement with wavelength. This may derive from a λ dispersion of the refractive index of the porous titania.

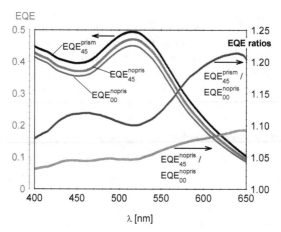

Fig. 25. Measured EQE for three simple configurations on a thin (3μm) DSC. The three configurations correspond to normal incidence (θ=00°) without prism, oblique incidence (θ=45°) without prism and oblique incidence (θ=45°) with prism. The light spot is always wholly inside the active area (impinging power is constant). The enhancement ratios are retrieved normalizing the last two curves to the first one.

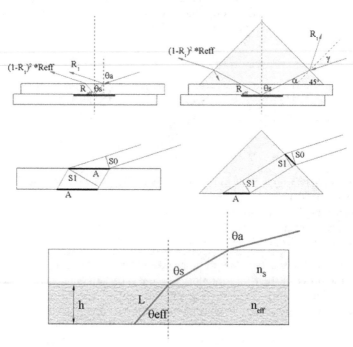

Fig. 26. Mechanism of the three main angular factors in the case of bare device and prism coupling. The considered effects are external reflectivity, projection of the active area and angular refractive path.

Now, the refractive term which we call ARP is not the only acting. Usually angular effects are known as detrimental on the output of photovoltaic devices. This is mainly due to a geometrical factor. When the cell is rotated in a wider light beam, the angular reduction of cross section seen by the cell reduces the collected power according to the *cosine law* (Balenzategui & Chenlo, 2005). Under a sun-simulator, such an effect does not allow to evaluate the other three angular factors we are interested in, since these are important at large angles but are screened by the convolution with the *cosine* term. With the laser, taking care that the spot size of the illuminating beam is always fully contained inside the active area, the impinging power is constant. Hence, the photocurrent measured for each incidence angle is directly proportional to the conversion efficiency and there is no need to take into account the *cosine law*. Instead, we will discuss in the following three other angular factors that are still acting on the photocurrent. These are the external air glass reflectance, the variation of the light intensity and the term of refraction ARP. A schematic of their action mechanism is drawn in the Fig. 26, for without and with coupling prism. The same external angle in air θ_a (considered between the light ray and the normal to the cell substrate) converts in two different angles of propagation inside the substrate θ_s for the bare DSC and with prism. Respectively: $\theta_s=asin(sin\theta_a/n_s)$; $\theta_s=45°-a=45°-asin(sin\gamma/n_s)= 45°-asin[sin(45°-\theta_a)/n_s]$. For simplicity we have used Snell law considering the same refractive index for the prism and the substrate n_s. At the same time, the reflection at their interface may be neglected. The first row of Fig. 26 represent the relevant angles and reflectance. What we consider here is the external air glass reflectance R_1 (in the figure) which varies with the incidence angle. The other one R_{eff} is the reflectance from the active layer (more precisely, multilayer stack). This can be used in attenuated total reflection (ATR) or different setups to study the internal layers but is not of interest here. In the following the external reflectance will be considered by using instead the term of transmittance $T(\theta)=1-R_1(\theta)$. The second row of the Fig. 26 represents the projection area A of the beam over the active area, respect to its external cross section S_0. The projection area affects, together with the transmittance, the light intensity over the titania $I(\theta)\propto T(\theta)/A(\theta)$. Finally, the last row represents the light path L inside the active layer which may be expressed using Snell law again, $L=h/cos\theta_{eff}=h/cos[asin(n_s sin\theta_s/n_{eff})]$. We are indicating the refractive index of the titania layer as an effective index n_{eff} since the nanoporous nature of the titania. It is anchored with dye molecules and its porosity filled with the electrolyte. Hence, according to a Bruggemann effective medium approximation, n_{eff} should be somewhat in between the index of a bulk titania (in anatase phase) and the electrolyte one (mainly due to its solvent).

To represent such angular factors for both bare DSCs and DSC plus prism, we prefer using the internal angle in glass, θ_s. The transmittance $T(\theta)$ term is well known by the Fresnel law which can be applied for both bare DSC and prism configurations. For the bare cell, T appears symmetrical when representing it versus both positive and negative angles of incidence ($-90°\leq\theta_a\leq+90°$ => $-40°<\theta_s<+40°$). The same doesn't hold for the prism and the same full θ_a range converts in an asymmetrical different θ_s range ($+05°<\theta_s<+85°$, when considering a single prism side as the entrance one). In such case, the T factor is a limiting one only for very larger positive angles. In the case of using an adequate emi-cylindrical prism (instead that the half cube one) T would be constant across a full range of θ_s. It should be noted that experimentally, the two ranges cannot be fully explored. At grazing angle the cross section of the substrate or the prism side becomes too small. The variation of the projection area has been represented as S_0/A, retrieved by means of geometrical considerations from the previous Fig. 26 and Snell law once more. In the case of the bare cell such factor follows a cosine law. Finally, we plot the angular refractive path as $L=h/\cos\theta_{eff}=h/\cos[asin(n_s sin\theta_s/n_{eff})]$ for different indexes of the titania layer. The lower the n_{eff} the larger the inclination of the rays, hence the lengthening of the path. The ARP factor depends on θ_{eff} and appears the same for the two configurations, when representing it towards the internal angle θ_s.

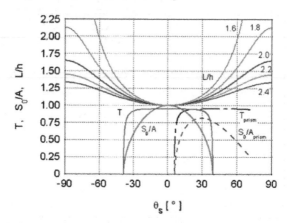

Fig. 27. Computed analytical forms of the three different angular factors that affect I_{SC}. Normalized projection section S_0/A of the beam over the titania (bare cell, bottom central, and with prism, bottom right). Transmitted light power T at the external air/glass interface without (middle central) and with prism (middle right), for the case of unpolarized light. Normalized light path L/h for five different effective indexes of the titania/electrolyte phase (top curves, from bottom to top n_{eff} = 2.4, 2.2, 2.0, 1.8 and 1.6).

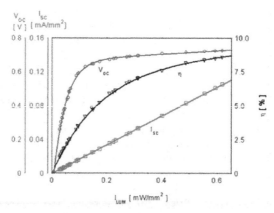

Fig. 28. Main parameters of a DSC measured versus the light intensity at λ=545nm and normal incidence. The I_{SC} could be fitted to a linear curve, while the V_{OC} and the power efficiency η to an exponential plus a linear. The light spot from a lamp was about 8mm^2.

In the Fig. 27 the computed variation of the three factor are represented. Using a simple approximation of a Lambert Beer exponential absorption in the sensitized titania, we may write the quantum efficiency as:

$$EQE = LHE \cdot IQE = T(\theta_s) \cdot [1 - e^{-\alpha L(\theta_s)}] \cdot \eta_{inj}\eta_{col} \qquad (7)$$

The effect of the second mechanism, the projection area and its influence on the intensity, may be supposed to potentially affect the internal quantum efficiency IQE via one or both of its subterms, the injection η_{inj} and collection η_{col} efficiencies (Trupke et al., 2000). Since the T factor is well known, and we are mainly interested in the refractive path, the effect of the intensity yet unknown needs to be quantified or cleared out in some other way. In the Fig. 28 we report measurements of the main parameters of a standard DSC (thick, h=12μm) towards the intensity I_{LUM} of a monochromatic beam (λ=545nm), at normal incidence (θ_s=0°). The short circuit current I_{SC} keeps linear in the full used range. Hence, in such range there is no apparent dependence of the EQE on the intensity. The following angular measurements were executed with a light intensity which remains inside the I_{LUM} range of Fig. 28.

The Fig. 29 presents angular measurements on thick and thin cells at a wavelength (λ=633nm) where EQE is quite low, about one third of the maximum. The two central curves are the measurements on a standard DSC (thick, h=12μm). The lowest of the two (M-shaped) represents the bare EQE while the upper one (U-shaped) is the EQE normalized to the transmittance T (top solid line, blue on line). As it can be observed the resulting normalized EQE is monotonically increasing with the module of the angle θ. It could be fitted by using the refractive term $1-e^{-\alpha h/\cos[a\sin(n_s\sin\theta_s/n_{eff})]}$ retrieving an effective index of n_{eff} = 2.21.

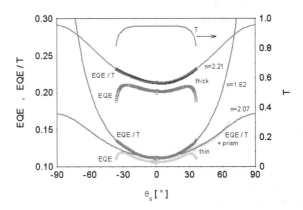

Fig. 29. EQE measured on thick and thin DSCs at λ=633nm. All normalized curves could be fitted to the angular path (Lambert-Beer as a function of the angle). The retrieved effective indexes depend on the specific cell and for a given cell also upon using or not the prism. This may be due to different TiO$_2$ porosities and thicknesses and to single cell disuniformity.

Experimentally, the symmetry of the curves indicate a proper alignment of the light beam with respect to the axis of rotation. The two bottom central curves represent similar measurements on a thin (h=3μm) cell. In this case, a larger angular enhancement could be seen, considered both in absolute and in percentage values. This is expected since the lower h. The fit with the refractive term is rather good, but the retrieved n_{eff} = 1.62 is much lower than the previous one on the thick cell. This could be ascribed to real differences in the titania layer porosity, but the difference appears to be too large. Another physical effect may be taken into consideration. Increasing the angle, and so the path lengthening, means to effectively absorb photons and generate charges closer to the photoelectrode. So the electrolyte ions should pass through a longer percolation path across the nanoporous titania. This supposedly influences the collection efficiency η_{col} in a different way between thick and thin cells. An outlook of this work is to investigate such potential effect with electrochemical impedance spectroscopy (EIS) measurements at different angles of incidence.

The bottom curve on the right is the normalized EQE when using the coupling prism on the thin cell. Also now we fitted (somewhat less well at large angles) with the refractive path, but the refractive index is still different (n_{eff} = 2.07) and this time on the same cell. Such aspect is not good and will require further investigations in the future, to reach a good agreement between measurements on the bare cell and the prism case. Experimentally, when excluding the intensity factor, a further issue could be still influencing. Indeed, if spatial disuniformity of the layer is present (hence, of EQE across the active area), the enlargement of the beam projection with the angle could affect the angular trend by sampling regions with different EQE. This could partially explain the observed differences and indicate the importance of having DSCs with a very uniform titania layer for the angular measurements. Another outlook for future implementation is the execution of measurements with also an emi-cylindrical prism and the correlation of the results obtained with all the three coupling elements (bare cell, half-cube prism, emi-cylindrical prism).

5. Conclusions

The technology of the dye solar cells offers nowadays a quite assessed typical configuration. Yet, different strategies are currently under study for increasing the stability and lifetime, but also for the improvement of the energy conversion efficiency. Among such techniques, acting on the main components of the DSCs remains almost a must issue. Traditionally, modifying the titania properties, looking for different fabrications or semiconductors, together with the research of new dyes and electrolytes, still seem a very wide action field, too wide to make sure previsions but surely promising in the actual energetic panorama. Besides, approaches derived from nano photonic and plasmonic technology are being integrated in the dye solar cells with more complex schemes to further improve efficiency in a wide sense of photon management. In such a context, the present work investigated the use of scattering layers, double dye co-sensitization, photonic crystals and also angular refractive path. Till now, the use of diffusive scattering layers led to the best performances. Typical size of the used opaque particles (150 to 400 nanometers) causes along all the wavelengths a reflectance of the non absorbed light back into the cell. Quantitatively the effect caused the improvement of a relative 47% in photocurrent and over 40% in efficiency. The co-sensitization technique was approached by using the conventional Red Dye together with the SDA1570 typically used for dye laser systems. The co-sensitization procedure was showed to be effective, putting into evidence non-linear effects by synergic mechanisms. The performance of co-sensitized cells outperform the sum of those with the single dyes. The immersion of the partially N719-sensitized photoanode in a SDA solution induces the saturation of the remaining free TiO_2 surface and at the same time a partial displacement of the already attached N719 molecules, organizing the two molecules in such a way which limits the energy losses due to the excitonic interactions between homologue molecules. Despite the lower efficiency (~20%) of the co-sensitized DSCs with respect to standard ones, transparency is gained (doubled), confirming as an effective strategy for BIPV applications. The integration of photonic crystals into the DSC for structural coloration represents one of the most engaging results. In fact, despite the color created by the silica/titania multilayer grown onto the nanocrystalline TiO_2 together with the color conferred by the dye itself, the cell is able to enhance the photocurrent and efficiency because of interferential reflections.

Variously colored DSCs were fabricated with the proper designs of the photonic crystals. The PCs schemes led to ca. 60% enhancement in efficiency and ca. 50% in Jsc, on thin DSCs, with 3 micrometers thick nc-TiO_2 titania. Also in this case the devices are quite transparent, conferring other important properties / instruments to DSC, i.e. for the building integration. The differences between SLs and PCs arrangements have to be ascribed to their basic issues. Photonic Crystals allow selective reflections based on their structural order, while SLs rely on a random configuration. Hence, PCs effect results in well defined enhancements corresponding to the reflected bands, whereas the typically used SLs lead to an increased absorption at the longer wavelengths with a tailoring at the shorter ones. Finally, the PCs are capable to confer a color independent from the dye color and mostly important keeping a good DSCs transparency; the SLs instead are affected by losing DSCs transparency at all. One main contribution of this work has been to realize and discuss state-of-the-art implementations of all these techniques, which actually are largely studied in the literature. Besides them, also an apparently minor feature was investigated responding to a very basic concept of photon management. The application of an angular setup to illuminate DSCs, allowed to quantify the response of the cells to oblique incidence of light. Apart from the power loss due to the reduction of cross section according to the cosine law, the IPCE rises. The bare cells present a maximum in IPCE at an angle of incidence in between 50° and 60°. This happens although the reflectance of the external air/glass interface grows with angle. Such a feature is ascribed to a photon path lengthening, i.e., an angular refractive path. Upon using a coupling prism two main advantages are obtained. The cut-off in the external transmittance can be overcome and at the same time larger internal angles can be achieved. A simplified yet robust model, based on Fresnel reflectance, Snell law of refraction and Lambert-Beer absorption is able to fit the angular dependencies of the quantum efficiency. The model depends on the effective refractive index of the mesoporous titania layer, which can be set as a fit parameter together with the optical absorbance. Hence, the angular IPCE allows also an investigation of the internal active layer, even though simplified at this stage. The enhancement is active due to the typical features of the thin absorbing titania layer, i.e., its refractive index and absorption coefficient, which offer margin improvements for a ray propagating obliquely to be more absorbed. Hence, it depends on the wavelength, thickness and porosity of the titania layer, but also, for example, on the electrolyte refractive index too. From a photon management point of view, the angular effect has in common with the SLs and PCs to be more effective on thin DSCs and at wavelength where they absorbs less. Unfortunately, its disadvantage is to suffer from the reduction of cross section and power. Nevertheless, it can still give suggestions on structuring DSCs working at normal incidence, and, it cannot be excluded also its potential use in proper new designs of thin films DSCs.

6. References

Abbotto, A.; Barolo, C.; Bellotto, L.; De Angelis, F.; Graetzel, M.; Manfredi, N.; Marinzi, C.; Fantacci, S.; Yum, J.-H. & Nazeeruddin, Md. K. (2008). Electron-rich heteroaromatic conjugated bipyridine based ruthenium sensitizer for efficient dye sensitized solar cells. *Chemical Communications* 42, 5318-5320, doi:10.1039/b900208a

Arakawa, H.; Yamaguchi, T.; Takeuchi, A. & Agatsuma, S. (2006). Efficiency Improvement of Dye-Sensitized Solar Cell by Light Confined Effect, *Conference Record of the IEEE 4th WCPEC*, 1, 36-39, ISBN 1-4244-0017-1, Waikoloa, Hawaii, USA, May 7-12, 2006, doi:10.1109/WCPEC.2006.279340

Arakawa, H.; Yamaguchi, T.; Takeuchi, A.; Okada, K.; Ezure, T. & Tanabe, N. (2007). Solar Cells Module. *JP2007012377 (A)*

Balenzategui, J. L. & Chenlo, F. (2005). Measurement and analysis of angular response of bare and encapsulated silicon solar cells. *Solar Energy Materials and Solar Cells* 86, 1, 53-83, doi:10.1016/j.solmat.2004.06.007

Barnes, P. R. F.; Anderson, A. Y.; Koops, S. E.; Durrant, J. R. & O'Regan, B. C. (2008). Electron injection efficiency and diffusion length in dye-sensitized solar cells derived from incident photon conversion efficiency measurements. *Journal of Physical Chemistry C* 113, 1126-1136

Bisquert, J.; Fabregat-Santiago, F.; Mora-Sero, I.; Garcia-Belmonte, G. & Gimenez, S. (2009). Electron Lifetime in Dye-Sensitized Solar Cells: Theory and Interpretation of Measurements. *Journal of Physical Chemistry C* 113, 40, 17278–17290, doi:10.1021/jp9037649

Buscaino, R.; Baiocchi, C.; Barolo, C.; Medana, C.; Graetzel, M., Nazeeruddin, Md. K. & Viscardi, G. (2007). A mass spectrometric analysis of sensitizer solution used for dye-sensitized solar cell. *Inorganica Chimica Acta* 361, 798–805

Calvo, M. E.; Colodrero, S.; Rojas, T. C.; Anta, J. A.; Ocaña, M. & Míguez, H. (2008). Photoconducting Bragg Mirrors based on TiO2 Nanoparticle Multilayers. *Advanced Functional Materials* 18, 18, 2708-2715, doi:10.1002/adfm.200800039

Catchpole, K. R. & Polman, A. (2008). Plasmonic solar cells. *Optics Express* 16, 26, 21793-21800, doi:10.1364/OE.16.021793

Chappel, S.; Grinis, L.; Ofir, A. & Zaban, A. (2005). Extending the Current Collector into the Nanoporous Matrix of Dye Sensitized Electrodes. *Journal of Physical Chemistry B* 109, 1643-1647, doi:10.1021/jp044949+

Chen, C.-Y.; Wu, S.-J.; Wu, C.-G.; Chen, J.-G.; Ho, K.-C. New ruthenium complexes containing oligoalkylthiophene-substituted 1,10-phenanthroline for nanocrystalline dye-sensitized solar cells. *Advanced Functional Materials* 17, 1, 29–36

Chen, D.; Huang, F.; Cheng, Y.-B. & Caruso, R. A. (2009). Mesoporous Anatase TiO2 Beads with High Surface Areas and Controllable Pore Sizes: A Superior Candidate for High-Performance Dye-Sensitized Solar Cells. *Advanced Materials* 21, 21, 2206–2210

Chen, X. & Mao, S. S. (2007). Titanium Dioxide Nanomaterials: Synthesis, Properties, Modifications, and Applications. *Chemical Reviews* 107, 7, 2891–2959, doi:10.1021/cr0500535

Chen, Y.; Zeng, Z.; Li, C.; Wang, W.; Wang X. & Zhang, B. (2005). Highly efficient co-sensitization of nanocrystalline TiO2 electrodes with plural organic dyes. *New Journal of Chemistry* 29, 773-776, doi:10.1039/b502725j

Cid, J.-J.; Yum, J.-H.; Jang, S.-R.; Nazeeruddin, M. K.; Martinez-Ferrero, E.; Palomares, E.; Ko, J.; Graetzel, M. & Torres, T. (2007). Molecular Cosensitization for Efficient Panchromatic Dye-Sensitized Solar Cells. *Angewandte Chemie International Edition* 46, 44, 8358–8362, doi:10.1002/anie.200703106

Clifford, J. N.; Palomares, E.; Nazeeruddin, M. K.; Thampi, R.; Grätzel, M. & Durrant, J. R. (2004). *Journal of American Chemical Society* 126, 18, 5670-5671, doi:10.1021/ja049705h

Colodrero, S.; Ocaña, M. & Míguez, H. (2008). Nanoparticle-based one-dimensional photonic crystals. *Langmuir* 24, 9, 4430-4434

Colodrero, S.; Mihi, A.; Anta, J. A.; Ocaña, M. & Míguez, H. (2009a). Experimental demonstration of the mechanism of light harvesting enhancement in Photonic-Crystal-Based Dye-Sensitized Solar Cells. *Journal of Physical Chemistry C* 113, 4, 1150-1154

Colodrero, S.; Mihi, A.; Haggman, L.; Ocaña, M.; Boschloo, G.; Hagfeldt, A. & Míguez, H. (2009b). Porous One-Dimensional Photonic Crystals Improve the Power-Conversion Efficiency of Dye-Sensitized Solar Cells. *Advanced Materials* 21, 7, 764-770

Colonna, D.; Dominici, L.; D'Ercole, D.; Brunetti, A.; Michelotti, F.; Brown, T. M.; Reale, A. & Di Carlo, A. (2010). *Superlattice and Microstructures* 47, 197-201

Colonna, D.; Colodrero, S.; Miguez, H.; Di Carlo, A. (2011). Introducing structural color in dye sensitized solar cells by using photonic nanostructures: interplay between conversion efficiency and optical properties. [in print]

Colonna, D.; Capogna, V.; Lembo, A.; Reale, A.; Brown, T. M. & Di Carlo, A. (2011). Efficient co-sensitization of nanocrystalline titania photo-anodes of Dye-Sensitized Solar Cells. [in print]

Dai, S.; Wang, K.; Weng, J.; Sui, Y.; Huang, Y.; Xiao, S.; Chen, S.; Hu, L.; Kong, F.; Pan, X.; Shi, C. & Guo, L. (2005). Design of DSC panel with efficiency more than 6%. *Solar Energy Materials and Solar Cells* 85, 447-455

Di Carlo, A.; Reale, A.; Brown, T. M.; Cecchetti, M.; Giordano, F.; Roma, G.; Liberatore, M.; Mirruzzo, V. & Conte, V. (2008). Smart Materials and Concepts for Photovoltaics: Dye Sensitized Solar Cells, In: *Smart Materials for Energy, Communications and Security*, I.A. Luk'yanchuk & D. Mezzane, (Ed.), 97-126, Springer, ISBN 978-1-4020-8795-0 (Print), 978-1-4020-8796-7 (Online), Dordrecht, The Netherlands, doi:10.1007/978-1-4020-8796-7

Ding, I-K.; Zhu, J.; Cai, W.; Moon, S.-J.; Cai, N.; Wang, P.; Zakeeruddin, S. M.; Graetzel, M.; Brongersma, M. L.; Cui, Y. & McGehee, M. D. (2011). Plasmonic Dye-Sensitized Solar Cells. *Advanced Energy Materials* 1, 52-57

Dominici, L.; Vesce, L.; Colonna, D.; Michelotti, F.; Brown, T. M.; Reale, A. & Di Carlo, A. (2010). Angular and prism coupling refractive enhancement in dye solar cells. *Applied Physics Letters* 96, 103302, doi:10.1063/1.3328097

Ezaki, S.; Gonda, I.; Okuyama, Y.; Takashima, A. & Furusaki, K. (2006). Dye Sensitized Solar Cells. *JP2006294423 (A)*

Ferber, J. & Luther, J. (1998). Computer simulations of light scattering and absorption in dye-sensitized solar cells. *Solar Energy Materials and Solar Cells* 54, 1-4, 265-275

Goldstein, J.; Yakupov, I. & Breen, B. (2010). Development of large area photovoltaic dye cells at 3GSolar. *Solar Energy Materials and Solar Cells* 94, 4, 638-641

Graetzel, M. (2005). Solar energy conversion by dye-sensitized photovoltaic cells. *Inorganic Chemistry* 44, 20, 6841-6851

Gutierrez-Tauste, D.; Zumeta, I.; Vigil, E.; Hernandez-Fenollosa, M. A.; Domenech, X. & Ayllon, J. A. (2005). New low-temperature preparation method of the TiO2 porous photoelectrode for dye-sensitized solar cells using UV irradiation. *Journal of Photochemistry and Photobiology A: Chemistry* 175, 165-171 doi:10.1016/j.jphotochem.2005.04.031

Hagfeldt, A. & Graetzel, M. (1995). Light-Induced Redox Reactions in Nanocrystalline Systems. *Chemical Reviews* 95, 1, 49-68, doi:10.1021/cr00033a003

Hägglund, C.; Zäch, M. & Kasemo, B. (2008). Enhanced charge carrier generation in dye sensitized solar cells by nanoparticle plasmons. *Applied Physics Letters* 92, 1, 013113, doi:10.1063/1.2830817

Halaoui, L. I.; Abrams, N. M. & Mallouk, T. (2005). Increasing the conversion efficiency of dye-sensitized TiO2 photoelectrochemical cells by coupling to photonic crystals. *Journal of Physical Chemistry B*, 109, 13, 6334-6342

Halme, J.; Boschloo, G.; Hagfeldt, A. & Lund, P. (2008). Spectral Characteristics of Light Harvesting, Electron Injection, and Steady-State Charge Collection in Pressed TiO2 Dye Solar Cells. *Journal of Physical Chemistry C* 112, 5623-5637, doi:10.1021/jp711245f

Hinsch, A.; Kroon, J. M.; Kern, R.; Uhlendorf, I.; Holzbock, J.; Meyer, A. & Ferber, J. Long-term stability of dye-sensitised solar cells (2001). *Progress in Photovoltaics: Research and Applications* 9, 6, 425–438, doi:10.1002/pip.397

Hinsch, A.; Behrens, S.; Berginc, M.; Bönnemann, H.; Brandt, H.; Drewitz, A.; Einsele, F.; Faßler, D.; Gerhard, D.; Gores, H.; Haag, R.; Herzig, T.; Himmler, S.; Khelashvili, G.; Koch, D.; Nazmutdinova, G.; Opara-Krasovec, U.; Putyra, P.; Rau, U.; Sastrawan, R.; Schauer, T.; Schreiner, C.; Sensfuss, S.; Siegers, C.; Skupien, K.; Wachter, P.; Walter, J.; Wasserscheid, P.; Würfel, U. & Zistler, M. (2008). Material Development for Dye Solar Modules: Results from an Integrated Approach. *Progress in Photovoltaics: Research and Applications* 16, 6, 489–501

Hore, S.; Nitz, P.; Vetter, C.; Prahl, C.; Niggemann, M. & Kern, R. (2005). Scattering spherical voids in nanocrystalline TiO2 - enhancement of efficiency in dye-sensitized solar cells. *Chemical Communications* 15, 2011-2013, doi:10.1039/B418658N

Hore, S.; Vetter, C.; Kern, R.; Smit, H. & Hinsch, A. (2006). Influence of scattering layers on efficiency of dye-sensitized solar cells. *Solar Energy Materials and Solar Cells* 90, 9, 1176-1188, doi:10.1016/j.solmat.2005.07.002

Ito, S.; Matsui, H.; Okada, K.; Kusano, S.; Kitamura, T.; Wada, Y. & Yanagida, S. (2004). Calibration of solar simulator for evaluation of dye-sensitized solar cells. *Solar Energy Materials and Solar Cells* 82, 3, 421-429, doi:10.1016/j.solmat.2004.01.030

Ito, S.; Nazeeruddin, M.; Liska, P.; Comte, P.; Charvet, R.; Pechy, P.; Jirousek, M.; Kay, A.; Zakeeruddin, S. & Graetzel, M. (2006). Photovoltaic Characterization of Dye-sensitized Solar Cells: Effect of Device Masking on Conversion Efficiency. *Progress in photovoltaics: research and applications* 14, 7, 589-601

Ito, S.; Chen, P.; Comte, P.; Nazeeruddin, M. K.; Liska, P.; Pechy, P. & Graetzel, M. (2007). Fabrication of screen-printing pastes from TiO2 powders for dye-sensitised solar cells. *Progress in photovoltaics: research and applications*, 15, 6, doi:10.1002/pip.768

Kalyanasundaram, K. & Graetzel, M. (1998). Applications of functionalized transition metal complexes in photonic and optoelectronic devices. *Coordination Chemistry Reviews* 177, 1, 347-414

Kay, A. (1995). Dye-stabilised photovoltaic cell module. *DE4416247 (A1)*

Khelashvili, G.; Behrens, S.; Weidenthaler, C.; Vetter, C.; Hinsch, A.; Kern, R.; Skupien, K.; Dinjus, E. & Bonnemann, H. (2006). Catalytic platinum layers for dye solar cells: A comparative study. *Thin Solid Films*, 511–512, 342–348, doi:10.1016/j.tsf.2005.12.059

Kim, C.; Lior, N. & Okuyama, K. (1996). Simple mathematical expressions for spectral extinction and scattering properties of small size-parameter particles, including examples for soot and TiO2. *Journal of Quantitative Spectroscopy and Radiative Transfer* 55, 3, 391-411, doi:10.1016/0022-4073(95)00160-3

Kim, H.; Auyeung, R. C. Y.; Ollinger, M.; Kushto, G. P.; Kafafi, Z. H. & Piqué, A. (2006). Laser-sintered mesoporous TiO2 electrodes for dye-sensitized solar cells. *Applied Physics A: Materials Science & Processing* 83, 1, 73-76, doi:10.1007/s00339-005-3449-0

Kim, S. S.; Nah, Y. C.; Noh, Y. Y.; Jo, J. & Kim, D. Y. (2006). Electrodeposited Pt for cost-efficient and flexible dye-sensitized solar cells. *Electrochimica Acta*, 51, 3814–3819, doi:10.1016/j.electacta.2005.10.047

Koide, N. & Han, L. (2004). Measuring methods of cell performance of dye-sensitized solar cells. *Review Of Scientific Instruments* 75, 9, 2828-2831, doi:10.1063/1.1784556

Koo, H. J.; Park, J.; Yoo, B.; Yoo, K.; Kim, K. & Park, N. G. (2008). Size-dependent scattering efficiency in dye-sensitized solar cell. *Inorganica Chimica Acta* 361, 3, 677-683

Koops, S. E. & Durrant, J. R. (2008). Transient emission studies of electron injection in dye sensitised solar cells. *Inorganica Chimica Acta* 361, 3, 663-670

Kroon, J. M.; Bakker, N. J.; Smit, H. J. P.; Liska, P.; Thampi, K. R.; Wang, P.; Zakeeruddin, S. M.; Grätzel, M.; Hinsch, A.; Hore, S.; Würfel, U.; Sastrawan, R.; Durrant, J. R.; Palomares, E.; Pettersson, H.; Gruszecki, T.; Walter, J.; Skupien, K. & Tulloch, G. E. (2007). Nanocrystalline dye-sensitized solar cells having maximum performance. Progress in Photovoltaics: Research and Applications 15, 1, 1-18, doi:10.1002/pip.707

Kuang, D.; Walter, P.; Nuesch, F.; Kim, S.; Ko, J.; Comte, P.; Zakeeruddin, S. M.; Nazeeruddin, M. K. & Graetzel, M. (2007). Co-sensitization of Organic Dyes for Efficient Ionic Liquid Electrolyte-Based Dye-Sensitized Solar Cells. Langmuir 23, 10906-10909, doi: 10.1021/la702411n

Lee, S.-H. A.; Abrams, N. M.; Hoertz, P. G.; Barber, G. D.; Halaoui, L. I. & Mallouk, T. E. (2008). Coupling of Titania Inverse Opals to Nanocrystalline Titania Layers in Dye-Sensitized Solar Cells. Journal of Physical Chemistry B 112, 46, 14415-14421

Lindström, H.; Magnusson, E.; Holmberg, A.; Södergren, S.; Lindquist, S.-E. & Hagfeldt, A. (2002). A new method for manufacturing nanostructured electrodes on glass substrates. Solar Energy Materials and Solar Cells 73, 91, doi:10.1016/S0927-0248(01)00114-3

Lozano, G.; Colodrero, S.; Caulier, O.; Calvo, M. E. & Miguez, H. (2010). Theoretical Analysis of the Performance of One-Dimensional Photonic Crystal-Based Dye-Sensitized Solar Cells. Journal of Physical Chemistry C 114, 8, 3681-3687

Meyer, T.; Martineau, D.; Azam, A. & Meyer, A. (2007). All Screen Printed Dye Solar Cell, Proceeding of SPIE 6656, 665608.1-665608.11, ISBN 978-0-8194-6804-8, San Diego, California, USA, August 28-30, 2007

Meyer, T.; Scott, M.; Martineau, D.; Meyer, A. & Cainaud, Y. (2009). Turning the dye solar cell into a product, 3rd International Conference Industrialization of DSC, DSC-IC 09, Nara, Japan, April 22-24, 2009

Mihi, A. & Miguez, H. (2005). Origin of Light-Harvesting Enhancement in Colloidal-Photonic-Crystal-Based Dye-Sensitized Solar Cells. Journal of Physical Chemistry B 109, 15968-15976

Mihi, A.; López-Alcaraz, F. J. & Míguez, H. (2006). Full spectrum enhancement of the light harvesting efficiency of dye sensitized solar cells by including colloidal photonic crystal multilayers. Applied Physics Letters 88, 193110

Mihi, A.; Calvo, M.; Anta, J. A. & Miguez, H. (2008). Spectral Response of Opal-Based Dye-Sensitized Solar Cells. Journal of Physical Chemistry C 112, 1, 13-17

Mincuzzi, G.; Vesce, L.; Reale, A.; Di Carlo, A. & Brown, T. M. (2009). Efficient sintering of nanocrystalline titanium dioxide films for dye solar cells via raster scanning laser Applied Physics Letters 95, 103312, doi:10.1063/1.3222915

Mincuzzi, G.; Vesce, L.; Liberatore, M.; Reale, A.; Di Carlo, A. & Brown, T. M. (2011). Dye Laser Sintered TiO_2 Films for Dye Solar Cells Fabrication: an Electrical, Morphological and Electron Lifetime Investigation. IEEE Transaction on Electron Devices, doi:10.1109/TED.2011.2160643 [in print]

Murakami, T. N. & Graetzel, M. (2008). Counter electrodes for DSC: Application of functional materials as catalysts. Inorganica Chimica Acta 361, 572-580

Nazeeruddin, M.; Kay, A.; Rodicio, I.; Humphry-Baker, R.; Muller, E.; Liska, P.; Vlachopoulos, N. & Graetzel, M. (1993). Conversion of light to electricity by cis-X2bis(2,2'-bipyridyl-4,4'-dicarboxylate)ruthenium(II) charge-transfer sensitizers (X = Cl-, Br-, I-, CN-, and SCN-) on nanocrystalline titanium dioxide electrodes. Journal of the American Chemical Society 115, 14, 6382, doi:10.1021/ja00067a063

Nazeeruddin, M. K.; De Angelis, F.; Fantacci, S.; Selloni, A.; Viscardi, G.; Liska, P.; Ito, S.; Takeru, B. & Graetzel, M. (2005). Combined Experimental and DFT-TDDFT

Computational Study of Photoelectrochemical Cell Ruthenium Sensitizers. *Journal of the American Chemical Society* 127, 16835-16847, doi:10.1021/ja0524671

Nishimura, S.; Abrams, N.; Lewis, B. A.; Halaoui, L. I.; Mallouk, T. E.; Benkstein, K. D.; Van de Lagemaat, J. & Frank, A. J. (2003). Standing wave enhancement of red absorbance and photocurrent in dye-sensitized titanium dioxide photoelectrodes coupled to photonic crystals. *Journal of the American Chemical Society* 125, 20, 6306-6310

Ogura, R. Y.; Nakane, S.; Morooka, M.; Orihashi, M.; Suzuki, Y. & Noda, K. (2009). High-performance dye-sensitized solar cell with a multiple dye system. *Applied Physics Letters* 94, 073308, doi:10.1063/1.3086891

Okada, K. & Tanabe, N. (2006). Dye-Sensitized Solar cell and its manufacturing method. *JP2006236960 (A)*

Ono, T.; Yamaguchi, T. & Arakawa, H. (2009). Study on dye-sensitized solar cell using novel infrared dye. *Solar Energy Materials and Solar Cells* 93, 831–835

O'Regan, B. & Graetzel, M. (1991). A low-cost, high-efficiency solar cell based on dye-sensitized colloidal TiO2 films. *Nature* 353, 737-740, doi:10.1038/353737a0

Pan, H.; Ko, S. H.; Misra, N. & Grigoropoulos, C. P. (2009). Laser annealed composite titanium dioxide electrodes for dye-sensitized solar cells on glass and plastics. *Applied Physics Letters* 94, 071117, doi:10.1063/1.3082095

Pala, R. A.; White, J.; Barnard, E.; Liu, J. & Brongersma, M. L. (2009). Design of Plasmonic Thin-Film Solar Cells with Broadband Absorption Enhancements. *Advanced Materials* 21, 1-6

Pandey, S. S.; Inoue, T.; Fujikawa, N.; Yamaguchi, Y. & Hayase, S. (2010). Substituent effect in direct ring functionalized squaraine dyes on near infra-red sensitization of nanocrystalline TiO2 for molecular photovoltaics. *Journal of Photochemistry and Photobiology A: Chemistry* 214, 2-3, 269-275

Park, N.-G. (2010). Light management in dye-sensitized solar cell. *Korean Journal of Chemical Engineering* 27, 2, 375-384, doi:10.2478/s11814-010-0112-z

Peter, L. M. & Wijayantha, K. G. U. (2000). Electron transport and back reaction in dye sensitised nanocrystalline photovoltaic cells. *Electrochimica Acta* 45, 28, 4543-4551, doi:10.1016/S0013-4686(00)00605-8

Pettersson, H.; Gruszecki, T.; Bernhard, R.; Häggman, L.; Gorlov, M.; Boschloo, G.; Edvinsson, T.; Kloo, L. & Hagfeldt A. (2007). The monolithic muticell: A tool for testing material components in dye-sensitized solar cells, *Progress in Photovoltaics: Research and Applications* 15, 113-121

Pichot, F.; Pitts, J. R. & Greg, B. A. (2000). Low-Temperature Sintering of TiO2 Colloids: Application to Flexible Dye-Sensitized Solar Cells. *Langmuir*, 16, 13, 5626–5630, doi:10.1021/la000095i

Polo, A. S.; Itokazu, M. K. & Murakami Iha, N. Y. (2004). Metal complex sensitizers in dye-sensitized solar cells. *Coordination Chemistry Reviews* 248, 1343-1361 doi:10.1016/j.ccr.2004.04.013

Rühle, S.; Greenwald, S.; Koren, E. & Zaban, A. (2008). Optical Waveguide Enhanced Photovoltaics. *Optics Express* 16, 26, 21801-21806, doi:10.1364/OE.16.021801

Saito, Y.; Kitamura, T.; Wada, Y. & Yanagida, S. (2002). Application of Poly(3,4-ethylendioxythiophene) to Counter Electrode in Dye-Sensitized Solar Cells. *Chemistry Letters* 31, 10, 1060, doi:10.1246/cl.2002.1060

Saito, Y.; Kubo, W.; Kitamura, T.; Wada, Y. & Yanagida, S. (2004). I-/I-3(-) redox reaction behavior on poly(3,4-ethylenedioxythiophene) counter electrode in dye-sensitized solar cells. *Journal of Photochemistry and Photobiology A: Chemistry* 164, 1-3, 153-157, doi:10.1016/j.jphotochem.2003.11.01

Sastrawan, R.; Beier, J.; Belledin, U.; Hemming, S.; Hinsch, A.; Kern, R.; Vetter, C.; Petrat, F. M.; Prodi-Schwab, A.; Lechner, P. & Hoffmann, W. (2006). A glass frit-sealed dye solar cell module with integrated series connections. *Solar Energy Materials and Solar Cells* 90, 11, 1680-1691

Schlichthorl, G.; Park, N. G. & Frank, A. J. (1999). Evaluation of the Charge-Collection Efficiency of Dye-Sensitized Nanocrystalline TiO2 Solar Cells. *Journal of Physical Chemistry B* 103, 5, 782-791, doi:10.1021/jp9831177

Seaman, C. H. (1982). Calibration of solar cells by the reference cell method – The spectral mismatch problem. *Solar Energy* 29, 4, 291-298, doi:10.1016/0038-092X(82)90244-4

Shah, A.; Torres, P.; Tscharner, R.; Wyrsch, N. & Keppner, H. (1999). Photovoltaic Technology: The Case for Thin-Film Solar Cells. *Science* 285, 5428, 692-698

Solaronix S.A., April 2011 prod. spec. retrieved from http://www.solaronix.com/products/

Somani, P. R.; Dionigi, C.; Murgia, M.; Palles, D.; Nozar, P. & Ruani, G. (2005). Solid-state dye PV cells using inverse opal TiO2 films. *Solar Energy Materials and Solar Cells* 87, 513–519

Sommeling, P. M.; Rieffe, H. C.; van Roosmalen, J. A. M.; Schönecker, A.; Kroon, J. M.; Wienke, J. A. & Hinsch, A. (2000). Spectral response and IV-characterization of dye-sensitized nanocrystalline TiO2 solar cells. *Solar Energy Materials and Solar Cells* 62, 4, 399-410, doi:10.1016/S0927-0248(00)00004-0

Sommeling, P. M.; Späth, M.; Smit, H. J. P.; Bakker, N. J. & Kroon, J. M. (2004). Long-term stability testing of dye-sensitized solar cells. *Journal of Photochemistry and Photobiology A: Chemistry* 164, 137–144

Suzuki, K.; Yamaguchi, M.; Kumagai, M. & Yanagida, S. (2003). Application of Carbon Nanotubes to Counter Electrodes of Dye-sensitized Solar Cells. *Chemistry Letters* 32, 1, 28-29

Trupke, T.; Würfe, P. & Uhlendorf, I. (2000). Dependence of the Photocurrent Conversion Efficiency of Dye-Sensitized Solar Cells on the Incident Light Intensity. *Journal of Physical Chemistry B*, 104, 48, 11484–11488, doi:10.1021/jp001392z

Tulloch, G. E. (2004). Light and energy-dye solar cells for the 21st century. *Journal of Photochemistry and Photobiology A: Chemistry* 164, 1-3, 209-219, doi:10.1016/j.jphotochem.2004.01.027

Tulloch, G. E. & Skryabin, I. L. (2006). Photoelectrochemical devices. *KR20060035598 (A)*

Tvingstedt, K.; Persson, N.-K.; Inganäs, O.; Rahachou, A. & Zozoulenko, I. V. (2007). Surface plasmon increase absorption in polymer photovoltaic cells. *Applied Physics Letters* 91, 113514, doi:10.1063/1.2782910

Tvingstedt, K.; Dal Zilio, S.; Inganäs, O. & Tormen, M. (2008). Trapping light with micro lenses in thin film organic photovoltaic cells. *Optics Express* 16, 26, 21608-21615

Uchida, S.; Tomiha, M.; Masaki, N.; Miyazawa, A. & Takizawa, H. (2004). Preparation of TiO2 nanocrystalline electrode for dye-sensitized solar cells by 28 GHz microwave irradiation. *Solar Energy Materials and Solar Cells* 81, 135–139, doi:10.1016/j.solmat.2003.08.020.

Usami, A. (1997). Theoretical study of application of multiple scattering of light to a dye-sensitized nanocrystalline photoelectrichemical cell. *Chemical Physics Letters* 277, 1-3, 105-108, doi:10.1016/S0009-2614(97)00878-6

Van de Lagemaat, J.; Park, N. G. & Frank, A. J. (2000). Comparison of Dye-Sensitized Rutile - and Anatase-Based TiO2 Solar Cells. *Journal of Physical Chemistry B* 104, 2044-2052, doi:10.1021/jp994365l

Vesce, L.; Riccitelli, R.; Soscia, G.; Brown, T. M.; Di Carlo, A. & Reale, A. (2010). Optimization of nanostructured titania photoanodes for dye-sensitized solar cells:

Study and experimentation of TiCl$_4$ treatment. *Journal of Non-Crystalline Solids* 356, 37-40, 1958-1961

Vesce, L.; Riccitelli, R.; Orabona, A.; Brown, T. M.; Di Carlo, A. & Reale, A. (2011). Fabrication of Spacer and Catalytic Layers in Monolithic Dye-sensitized Solar Cells. *Progress in Photovoltaics: Research and Applications* [in print]

Walker, A. B.; Peter, L. M.; Lobato, K. & Cameron, P. J. (2006). Analysis of photovoltage decay transients in dye-sensitized solar cells. *Journal of Physical Chemistry B* 110, 50, 25504-25507

Wang, L.; Fang, X. & Zhang, Z. (2010). Design methods for large scale dye-sensitized solar modules and the progress of stability research. *Renewable and Sustainable Energy Reviews* 14, 3178-3184

Wang, P.; Zakeeruddin, M.; Moser, J. E.; Nazeeruddin, M. K.; Sekiguchi, T. & Graetzel, M. (2003). A stable quasi-solid-state dye sensitized solar cell with an amphiphilic ruthenium sensitizer and polymer gel electrolyte. *Nature Materials* 2, 402-407

Wang, Q.; Moser, J. E. & Graetzel, M. (2005). Electrochemical impedance spectroscopic analysis of dye-sensitized solar cells. *Journal of Physical Chemistry B* 109, 31, 14945-14953

Wang, Z. S.; Yamaguchi, T.; Sugihara, H. & Arakawa, H. (2005). Significant Efficiency Improvement of the Black Dye-Sensitized Solar Cell through Protonation of TiO2 Films. *Langmuir*, 21, 4272-4276, doi:10.1021/la050134w

Watson, T.; Mabbett, I.; Wang, H.; Peter, L. & Worsley, D. (2010). Ultrafast near infrared sintering of TiO2 layers on metal substrates for dye-sensitized solar cells. *Progress in Photovoltaics: Research and Applications*, [in print], doi:10.1002/pip.1041

Wijnhoven, J. E. G. J. & Vos, W. L. (1998). Preparation of Photonic Crystals Made of Air Spheres in Titania. *Science* 281, 5378, 802-804, doi:10.1126/science.281.5378.802

Yao, Q.-H.; Meng, F.-S.; Li, F.-Y.; Tian, H. & Huang, C.-H. (2003). Photoelectric conversion properties of four novel carboxylated hemicyanine dyes on TiO2 electrode. *Journal of Materials Chemistry* 13, 5, 1048–1053, doi:10.1039/B300083B

Yip, C.-H.; Chiang, Y.-M. & Wong, C.-C. (2008). Dielectric Band Edge Enhancement of Energy Conversion Efficiency in Photonic Crystal Dye-Sensitized Solar Cell. *Journal of Physical Chemistry C* 112, 23, 8735-8740

Yum, J-H.; Walter, P.; Huber, S.; Rentsch, D.; Geiger, T.; Nuesch, F.; De Angelis, F.; Graetzel, M. & Nazeeruddin, M. K. (2007). Efficient Far Red Sensitization of Nanocrystalline TiO2 Films by an Unsymmetrical Squaraine Dye. *Journal of American Chemical Society* 129, 34, 10320-10321, doi:10.1021/ja0731470

Yum, J.-H.; Moon, S. J.; Humphry-Baker, R.; Walter, P.; Geiger, T.; Nuesch, F.; Graetzel, M. & Nazeeruddin, M. K. (2008). Effect of coadsorbent on the photovoltaic performance of squaraine sensitized nanocrystalline solar cells. *Nanotechnology* 19, 42, 424005, doi:10.1088/0957-4484/19/42/424005

Zhang, D.; Yoshida, T. & Minoura, H. (2003). Low-Temperature Fabrication of Efficient Porous Titania Photoelectrodes by Hydrothermal Crystallization at the Solid/Gas Interface. *Advanced Materials* 15, 10, 814-817, doi:10.1002/adma.200304561

Zhang, Q. F.; Chou, T. P.; Russo, B.; Jenekhe, S. A. & Cao, G. Z. (2008). Polydisperse Aggregates of ZnO Nanocrystallites: A Method for Energy-Conversion-Efficiency Enhancement in Dye-Sensitized Solar Cells. *Advanced Functional Materials* 18, 11, 1654-1660, doi:10.1002/adfm.200701073

Zhou, Y.; Zhang, F.; Tvingstedt, K.; Tian, W. & Inganäs, O. (2008). Multifolded polymer solar cells on flexible substrates. *Applied Physics Letters* 93, 033302, doi:10.1063/1.2957995

3

Photo-Induced Electron Transfer from Dye or Quantum Dot to TiO₂ Nanoparticles at Single Molecule Level

King-Chuen Lin and Chun-Li Chang
Department of Chemistry, National Taiwan University, Taipei 106,
Institute of Atomic and Molecular Sciences, Academia Sinica, Taipei 106,
Taiwan

1. Introduction

Dye-sensitized solar cell (DSSC) has attracted wide attention for the potential application to convert sunlight into electricity. Organic dyes blended with TiO_2 nanoparticles (NPs) have been recognized as important light harvesting materials especially in the visible spectral range (Hara et al., 2002; Bisquert et al., 2002; Gratzel, 2001; Ferrere & Gregg, 2001; Hagfeldt & Gratzel, 2000; Cahen et al., 2000). The functional materials are assembled in a sandwiched type to undergo photon-induced current process. Following photoexcitation, the embedded dye molecules may lead to electron transfer (ET) to the TiO_2 conduction band. The injected electron flows through the semiconductor network and the external load to the counter electrode. At the counter electrode, the oxidized dye is reduced by electron donation from an electrolyte, and then the circuit becomes complete. The electron transfer kinetics in most dye/TiO_2 systems is as rapid as in the time regime of femtosecond to several hundred picoseconds. The injected electrons are localized in either subband or surface states of TiO_2 semiconductor. A fraction of the electrons, detrapped thermally from the reduced semiconductor, may possibly undergo recombination with the oxidized dye molecules. Such a back ET process takes place slowly from subnanoseconds to several milliseconds. An efficient solar cell deign should control lowering the rate of back electron transfer to prolong the lifetimes of charge-separated states. Therefore, characterizing kinetics of the forward and backward ET may be conducive to facilitating the working efficiency of a solar cell design.

Among a variety of DSSC designs, Grätzel and coworkers have applied ruthenium-based dyes adsorbed on the TiO_2 thin film, thereby achieving a very high power-conversion efficiency >11% (Gratzel, 2003, 2005). Despite a much lower efficiency in comparison, quantum dots (QDs) adopted recently to substitute for dyes have been popularly investigated including PbS (Plass et al., 2002; Ju et al., 2010), InAs (Yu et al., 2006), CdSe (Lee & Lo, 2009; Fan et al., 2010), CdS (Baker & Kamat, 2009; Lee & Lo, 2009), and PbSe (Luther et al., 2008; Choi et al., 2009). QDs have potential to be an alternative as electron donors (Robel et al., 2006; Kamat, 2008), for their unique properties such as size-dependent tunable energy gap (Yu et al., 2003; Kamat, 2008), a broad absorption band with large absorption cross sections (Yu et al., 2003), and multiple exciton generation (Yu et al., 2003; Luther et al., 2007; Kim et al., 2008; Sambur et al., 2010). When QDs absorb a photon to form an electron-hole pair, the electrons may have chance to transfer to an accepting species such as TiO_2, if the conduction band edge of QDs is tuned higher than the conduction band of TiO_2 (Robel et al., 2006; Yu et al., 2006; Kamat, 2008). Like the DSSC mechanism, kinetic behavior of ET between QDs and TiO_2 is one of the key roles to achieve a high energy-conversion efficiency. The bulk measurements yield ensemble-averaged information and sometimes could mask or overlook specific phenomena occurring at the sensitizer-semiconductor interfaces. For instance, conformation change and reorientation of the adsorbate structure feasibly induce the fluctuation of fluorescence decay times for ET processes, but such detailed dynamical complexity can not be visualized in ensemble experiments (Moerner & Fromm, 2003; Michalet et al., 2006). As a result, single molecule spectroscopy (SMS) has emerged as a powerful tool for investigating the dynamic processes of excited molecules in heterogeneous surrounding (Xie & Dunn, 1994; Garcia-Parajo et al., 2000; Moerner & Fromm, 2003; Michalet et al., 2006; Gaiduk et al., 2007). Analysis of fluorescence intermittency observed in SMS, or called on/off blinking phenomena, has been widely studied to unveil the dynamic behaviors of triplet state (Yip et al., 1998; Veerman et al., 1999; Kohn et al., 2002), molecular reorientation (Ambrose et al., 1994), energy transfer (VandenBout et al., 1997; Cotlet et al., 2005; Flors et al., 2007), spectral diffusions (Xie & Dunn, 1994; Kulzer et al., 1997), and electron transfer (Holman & Adams, 2004; Bell et al., 2006; Wang et al., 2009; Chen et al., 2010). This chapter is confined to understand ET dynamics of either dyes or QDs adsorbed on the TiO_2 NPs thin film at a single molecule level.

By taking advantage of the SMS merits, the IFET phenomena of oxazine 1 dye and CdSe/ZnS core/shell QD adsorbed individually on the TiO_2 NPs surface are demonstrated. We acquired fluorescence trajectories of single dye molecule, characteristic of "on" and "off" blinking with time, in which the intensity fluctuation is attributed to the interfacial electron transfer (IFET) behavior. The fluorescent dye molecule lies in an off-blinking state when its electron moves to the TiO_2 NPs conduction band. While the electron transfers back to the oxidized state, the dye molecule blinks on again following photoexcitation. Discrete fluorescence intensity jumps between "on" and "off" level are analyzed by using autocorrelation function. The "on" and "off" lifetimes and the subsequent rate constants of forward and backward ET are then determined. On the other hand, the fluorescence lifetimes of QDs are measured as an alternative determination for the ET rate constants. QDs with different sizes are adopted for demonstration. We apply time-correlated single-photon counting (TCSPC) to measure fluorescence lifetimes among a quantity of QDs for each size. The fluorescence lifetime becomes shorter and the resultant "off" time is prolonged with decreased size of QDs. The off-time and on-time probability densities are then estimated and fitted appropriately. With the aid of Marcus model, theoretical ET rate constants are calculated for comparison and the ET process may thus be gained insight.

2. Experimental

2.1 Apparatus

All single molecule experiments were performed with a confocal fluorescence microscope. For the single dye (or QD) experiments, a single-mode pulsed laser at 630 nm (or 375 nm), with a repetition rate of 10 MHz (or 5 MHz) and pulsed duration of 280 ps (or 300 ps), was used as the excitation source. The beam collimated with a pair of lenses was spectrally filtered with an excitation filter before entering an inversed microscope. An oil immersion objective (100x, NA1.40) was used both to focus the laser beam onto the sample, prepared on the surface of a glass coverslip, and to collect fluorescence from the sample. The excitation intensity of the pulsed beam was constantly measured to be 40 - 210 W/cm^2 right on the top of the bare coverslip throughout the experiments. The fluorescence, after transmitting through a dichroic mirror, was refocused by a tube lens (200 mm focal length) onto an optical fiber (62.5 μm diameter) which was coupled to an avalanche photodiode (APD) detector with a 175 μm active area. Here the fiber serves as a pinhole to reject out-of-focus light. The fluorescence signal may also be reflected simultaneously to a charge-coupled device (CCD) by a beamsplitter. A notch filter (6<OD) or a combination of bandpass filters were positioned in front of the detector to remove excitation background. Given the wide-field images with a CCD detector, each fluorescent single molecule was readily moved to an illuminated position using a x-y positioning stage. The fluorescence lifetime of single molecule was measured by TCSPC with a TimeHarp 200 PCI-board (PicoQuant). The data were stored in a time-tagged time-resolved mode, which allowed recording every detected photon and its individual timing information. By taking into account deconvolution of the instrument response function (SymPhoTime by PicoQuant), the TCSPC curve was analyzed by single exponential tail-fit.

2.2 Material preparation

The TiO$_2$ precursor was prepared by a sol-gel process. 72 mL of 98% Titanium(IV) isopropoxide was added to 430 mL of 0.1 M nitric acid solution which was then stirred and heated to 85 ^{o}C for 8 h. The colloid, as filtered from the cooled mixture, was heated again in an autoclave at five temperatures of 180, 200, 220, 240, and 260 ^{o}C for 12 h to grow TiO$_2$ NPs. The TiO$_2$ colloid was concentrated to 13 wt%, followed by addition of a 30 wt% (with respect to TiO$_2$) of polyethylene glycol to prevent from cracking during drying.

To prepare for the thin film, 40-50 μL of 0.65% TiO$_2$ NPs aliquot was spin-coated on a cleaned coverslip. The same process was repeated four times to ensure TiO$_2$ well coated over the coverslip. After drying in the air for 30 min, the film was heated to 450 ^{o}C at a rate of 20 ^{o}C/min, and remained for 30 min before cooling to the room temperature. The phase and the size of TiO$_2$ NPs were characterized by using scanning electron microscopy. A drop of 30 μL of 0.1 nM oxazine 1 dye in methanol solution or 35 μL of 4 - 200 pg/L QDs in toluene solution was spin-coated over the TiO$_2$ NPs surface, and then put on the sample stage of microscope for fluorescence measurement.

3. Fluorescence intermittency and electron transfer by single dye molecule

3.1 Fluorescence lifetimes

As shown in Fig.1, the TiO$_2$ NPs thin film was well covered over the coverslip to ensure that the oxazine 1 dye may be fully adsorbed on the TiO$_2$ surfaces. The enlarged images displayed the NPs size about ~20 nm. Fig.2 shows absorption spectra of oxazine 1 solution in the presence (or absence) of TiO$_2$ NPs by means of ensemble measurements. The absorption cross section reaches as large as ~3.5x10^{-16} cm^2 at 630 nm. The fluorescence yield is 0.11 in ethanol as the solvent and increases to 0.19 in ethylene glycol (Sens & Drexhage, 1981). When the dye in methanol solution is diluted from 10^{-6} to 10^{-8} M, the spectral profile remains the same with a major peak and a minor shoulder. The spectra appear almost the same between the conditions with and without TiO$_2$ NPs involvement. The results suggest that the dimer contribution should be minor and negligible.

Fig. 1. Images of TiO_2 NPs thin film by using scanning electron microscopy. The average sizes of TiO_2 NP are about 20 nm for both films.

A 10^{-10} M of oxazine 1 dye solution was spin-coated on a bare or a TiO_2 NPs-coated coverslip. Fig.3(a) displays the fluorescence images of the single oxazine 1 molecules within a 24 μm x 24 μm area on the bare coverslip, as excited by a pulsed laser at 630 nm with excitation intensity of 39±2 W/cm². Each bright spot is attributed to a single molecular fluorescence. The diffraction-limited spot size is about ~650 nm. Fig.3(b) shows the background emissions of the TiO_2 NPs surface. Stray light scattered by the TiO_2 NPs causes background noise, which is however dimmer than those fluorescent spots from the dye molecules. Fig.3(c) shows the single molecular fluorescence images of dye molecules on TiO_2 NPs surface. The maximum fluorescence intensity reaches 650 counts acquired by a CCD within an integration time of 0.5 s. The fluorescent spots have different intensities resulting from variation of molecular orientation and micro-environments. The surface densities of fluorescent oxazine 1 on glass or TiO_2 coverslip are counted to be 0.088 or 0.085 per μm², respectively, with an uncertainty of ±10%. It is difficult to make a comparison with the surface density prepared initially, since a great amount of dye solution is sputtered away during the spin-coating process. The roughness of the TiO_2 NPs-coated coverslip may reduce the dye molecules from sputtering away to some extents, as compared to a smooth bare glass. Although the surface density of fluorescent spots is similar between them, some molecules in the excitation might be quenched by a faster IFET process before fluorescing and thus fail to be detected. The clefts or interstices in the TiO_2 film are probably ideal sites for effective IFET.

As shown in Fig.4(a), fluorescence decay of a single dye molecule on TiO_2 is obtained for the lifetime measurements by the TCSPC method. The decay curve is fitted to a single exponential function yielding a lifetime of 2.6 ns. In this manner, the lifetimes successively determined among 100 different oxazine 1 molecules are distributed in Fig.4(b), with an average value of 2.86±0.31 ns. The lifetime scattering reveals an inhomogeneous character among the dye molecules measured, because of variation of dipole orientation, transition frequency, and molecular polarization on the surface. In addition, the fluctuation of the lifetime is partially attributed to the power density variation which is about ±3%. For comparison, the average lifetimes over 100 oxazine 1 molecules are measured analogously to be 3.02±0.61 ns on the bare coverslip.

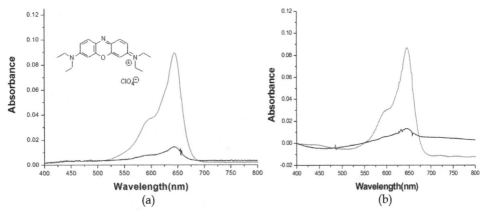

(a)
(b)

Fig. 2. (a) Absorption spectra of oxazine 1 dye in methanol solution with concentration of 1x10^{-6} (larger peak) and 1x10^{-8} M, respectively. Oxazine 1 structure is also displayed. (b) Absorption spectra of oxazine 1 methanol solution with concentration of 1x10^{-6} (larger peak) and 1x10^{-8} M, respectively, in the presence of TiO$_2$ NPs

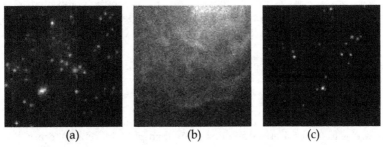

(a)　　　　　　　　(b)　　　　　　　　(c)

Fig. 3. Fluorescence images of (a) oxazine 1 molecules on bare coverslip, (b) TiO$_2$ NPs film without dye molecules involved, and (c) oxazine 1 molecules on TiO$_2$ NPs–coated coverslip.

Despite no statistical difference with the uncertainty considered, the average lifetime of the dye on the TiO$_2$ film is significantly smaller than that on the bare coverslip. While inspecting Fig.4(b), the asymmetric distribution shows more molecules lying on the side of shorter fluorescence lifetime. It suggests two points. First, some molecules might undergo even faster IFET process such that the excited state lifetimes become too short to be detected. Second, slight difference of the fluorescence lifetimes implies that most IFET processes should be slow and inefficient. The oxazine 1 molecule lacks effective anchoring groups like carboxylate in metal-polypyridine complexes to covalently bind the dye to the semiconductor (Oregan & Gratzel, 1991; Ramakrishna et al., 2005). These dye molecules randomly selected are anticipated to be physisorbed to the TiO$_2$ NPs surface or loosely trapped at the NPs interstices. However, that does not mean all interactions based on physisorption can not lead to efficient electron transfer. Some organic dyes such as cyanine or xanthene dyes incorporated in Langmuir-Blodgett films deposited on In$_2$O$_3$ or SnO$_2$ electrodes show very efficient electron transfer even without chemisorptions, due to a large free-energy difference between the lowest unoccupied molecular orbital (LUMO) of the dye and the edge of the conduction band of the semiconductor(Arden & Fromherz, 1980; Biesmans et al., 1991; Biesmans et al., 1992).

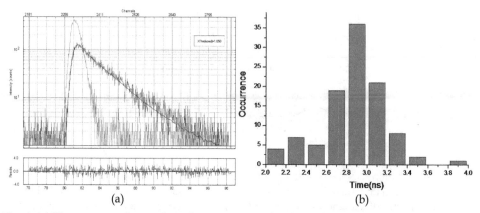

(a)   (b)

Fig. 4. (a) Fluorescence decay of single oxazine 1 molecule in the presence of TiO_2 NPs measured by using time-correlated single photon counting. The instrument response function gives a faster decay estimated to be 550 ps. (b) Distribution of excited state lifetimes among 100 single oxazine 1 molecules on the TiO_2 NPs surface, yielding an average lifetime of about 2.86 ns.

3.2 Fluorescence intensity trajectories

Fig.5(a) displays fluorescence intensity as a function of running time of a single oxazine 1 molecule on the bare coverslip with a 20 ms binning time to record the photon counts of emission. These time scales average out any faster fluorescence fluctuation or blinking of the dye molecules. The single-molecule trace shows a constant level of fluorescence intensity which then drops to the background in one step when photobleaching occurs. Wilkinson et al. reported the triplet state lifetime to be 14.5 µs for oxazine 1 in acetonitrile solution (Wilkinson et al., 1991), and a significantly longer lifetime is yet expected for the dye adsorbed on the solid film. A constant level of fluorescence intensity in Fig.5(a) implies that the prolonged lifetimes should be much shorter than the integration time adopted such that the triplet blinking may be smeared out. When 100 single molecules are sampled, about 70% display the trajectories like Fig.5(a). The remaining show blinks occasionally before photobleaching, in part because of photo-induced electron transfer to impurity sites of the glass coverslip. It is difficult to observe clearly triplet blinking of the single molecule caused by intersystem crossing. Instead, an average intensity of 5 counts/20 ms appears. The triplet state lifetime if assumed to be 14.5 µs amounts to 0.003 count, which is too weak to be resolved temporally.

As shown in Fig.6A(a), when the single-molecule fluorescence intensity trajectory of the dye molecules on the bare coverslip is slotted within 1 s-window, the fluorescence decay lifetimes are measured to be 3.63 ns and 3.59 ns for the time period of 54~55s and 266~267s, respectively (Fig.6A(b)). Alternatively, the lifetime fluctuation trajectory binned in 0.5 s-window gives rise to an average lifetime of 3.44±0.48 ns (Fig.6A(c)), which is consistent with that averaged over 100 molecules. These facts indicate that the fluorescence intensity fluctuation for oxazine 1 on glass is dominated by the radiative relaxation process from the singlet state.

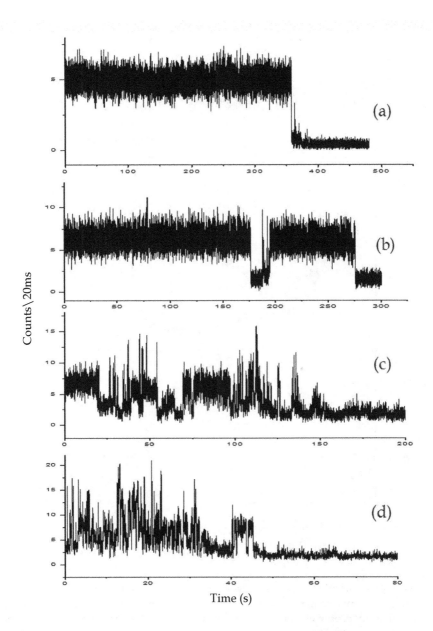

Fig. 5. Fluorescence trajectories recorded for single oxazine 1 molecules (a) on bare coverslip, and (b-d) on TiO$_2$ NPs-coated coverslip.

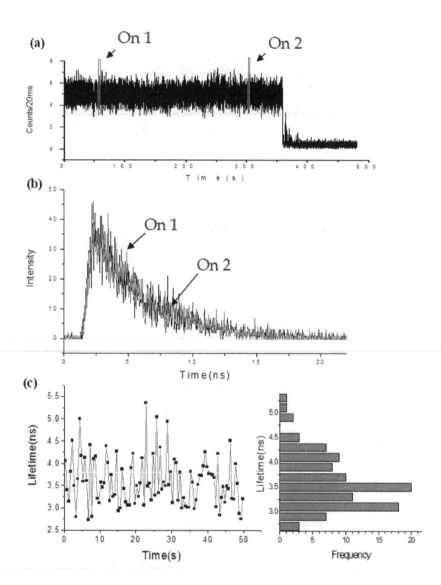

Fig. 6. (A) Single-molecule fluorescence decay profile of oxazine1 on bare glass. (a) The fluorescence intensity fluctuation slotted at 1 and 2 positions each within 1-s window. (b) Fluorescence decay profile for the 1 and 2 slots gives rise to the decay time of 3.63 ns and 3.59 ns, respectively. (c) The lifetime (binning window 0.5s) fluctuation trajectory (left), and the histogram for the lifetime trajectory from 0 to 50 s before photobleaching. The average lifetime is 3.44±0.48 ns.

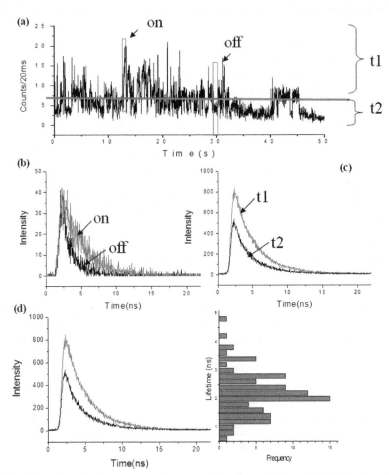

Fig. 6. (B) Single-molecule fluorescence decay profile of oxazine1 on TiO$_2$ NPs film. (a) The fluorescence intensity fluctuation slotted at one "on" position and the other "off" position. (b) Fluorescence decay profile for the "on" and "off" slots gives rise to the decay time of 2.93 and 1.26 ns, respectively. (c) Given a threshold set at 7 photocounts/20 ms, the emission trajectory is separated to higher level and lower level. Fluorescence decay profiles for high and low intensity have similar results as the "on" and "off" slots in (b). (d) The lifetime (binning window 0.5s) fluctuation in a range from 0.6 to 4.8 ns (left), and the histogram for the lifetime trajectory from 0 to 50 s before photobleaching.

On the other hand, Fig.5(b-d) show the fluorescence trajectories of individual dye molecules in the presence of TiO$_2$ NPs thin film ranging from several microseconds to seconds before photobleaching. These trajectories are classified into three types among 100 single dye molecules as sampled. About 20% of molecules yield the traces as in Fig.5(b), in which the interaction with TiO$_2$ NPs is weak such that the molecule fluoresces at most time. A 60% majority show the traces as in Fig.5(c), in which a stronger interaction is found and the "off" time becomes longer. The remaining 20% give the traces as in Fig.5(d), in which the molecule stays longer at "off" time. The molecule in Fig.5(d) should have relatively active electron transfer such that the fluorescence process is suppressed.

Fig.5(d) is used as an example. The fluorescence intensity trajectory is slotted within a 500-photon binned window to select one "on" intensity and the other "off" intensity (Fig.6B(a)). Analyzing the fluorescence decay yields a result of 2.93 and 1.26 ns for the "on" (12.85~13.33 s slot) and "off" (29.15~30.20 s slot) lifetime, respectively (Fig.6B(b)). Given a threshold at 7 counts/20 ms, the fluorescence intensity is divided to higher level and lower level. The lifetime analysis of these two levels yields the results similar to those obtained in the above time slots. The "on" state shows a twofold longer lifetime than the "off" state (Fig.6B(c)). This fact indicates that the fluorescence intensity fluctuation is caused by both factors of reactivity, i.e., the fraction of IFET occurrence frequency (Wang et al., 2009), and rate of electron transfer. The fluorescence lifetimes analyzed within 0.5s-window fluctuate in a range from 0.6 to 4.8 ns, which is more widely scattered than those acquired on the bare glass (Fig.6B(d)). This phenomenon suggests existence of additional depopulation pathway which is ascribed to ET between oxazine1 and TiO_2. However, other contribution such as rotational and translational motion of the dye on the TiO_2 film can not be rule out without information of polarization dependence of the fluorescence.

3.3 Autocorrelation analysis

An autocorrelation function based on the fluorescence intensity trajectory is further analyzed. When the dye molecules are adsorbed on the TiO_2 NPs surface, a four-level energy scheme is formed including singlet ground, singlet excited, and triplet states of the dye molecule as well as conduction band of TiO_2. Upon irradiation with a laser source, the excited population may undergo various deactivation processes. Because the selected dye molecule has a relatively short triplet excursion, the fluorescence in the absence of TiO_2 film becomes a constant average intensity with near shot-noise-limited fluctuation, as displayed in Fig.5(a) (Haase et al., 2004; Holman & Adams, 2004). As a result, the system can be simplified to a three-level energy scheme.

As the ET process occurs, the fluorescence appears to blink on and off. The transition between the on and off states may be considered as feeding between the singlet and the conduction subspaces (Yip et al., 1998),

$$\text{On} \underset{k_{on}}{\overset{k_{off}}{\rightleftharpoons}} \text{off} . \tag{1}$$

The on-state rate constant is equivalent to the backward ET rate constant from the conduction band, i.e.,

$$k_{on} = k_{bet} , \tag{2}$$

while the off-state rate constant corresponds to the excitation rate constant k_{ex} multiplied by the fraction of population relaxing to the conduction band, as expressed by

$$k_{off} = \frac{k_{et}}{k_{21} + k_{et}} k_{ex} . \tag{3}$$

Here, k_{21} is the relaxation rate constant from the excited singlet to ground state containing the radiative and non-radiative processes and k_{et} is the forward ET rate constant. k_{ex} is related to the excitation intensity I_o (units of erg/cm^2 s) by

$$k_{ex}=\sigma I_0/h\upsilon, \tag{4}$$

where σ is the absorption cross section and $h\upsilon$ is the photon energy. The average residence times in the on and off states correspond to the reciprocal of the feeding rate in the off and on states, respectively. That is, $\tau_{on} = 1/k_{off}$ and $\tau_{off} = 1/k_{on}$.

The rate constants in on-off transition may then be quantified by analyzing autocorrelation of fluorescence intensities (Holman & Adams, 2004). The normalized autocorrelation function is defined as the rate of detecting pairs of photons separated in time by an interval τ, relative to the rate when the photons are uncorrelated. It is expressed as

$$G(\tau) = \frac{< I(t)I(t+\tau) >}{< I(t) >^2}, \tag{5}$$

where I(t) is the fluorescence intensity at time t and τ is the correlation time. The bracketed term denotes the intensity average over time. When the population relaxation is dominated by the singlet decay, the autocorrelation function may be simplified to an exponential decay, i.e.,

$$G(\tau) = A + Be^{-k\tau}, \tag{6}$$

where A is an offset constant, B a pre-exponential factor, and k the decay rate constant. They are determined by fitting to the autocorrelation data. These parameters are explicitly related to the phenomena of on/off blinking due to the ET processes by,

$$k = k_{on} + k_{off} \tag{7}$$

and

$$\frac{B}{A} = \frac{k_{on}k_{off}(I_{on} - I_{off})^2}{(k_{on}I_{on} + k_{off}I_{off})^2}. \tag{8}$$

If $I_{on} >> I_{off}$, then the above equation is simplified to

$$\frac{B}{A} = \frac{k_{off}}{k_{on}}. \tag{9}$$

The forward and backward ET rate constants in the dye molecule-TiO$_2$ NPs system can thus be evaluated.

According to eq.5, Fig.7(a) shows that the autocorrelation result based on the fluorescence trajectory of the dye on glass (Fig.5(a)) appears to be noisy ranging from zero to microseconds. The dynamic information of the triplet state can not be resolved, consistent with the analyzed results of fluorescence decay times. When the dye molecule is on TiO$_2$, the fluorescence trajectory given in Fig.5(c) is adopted as an example for evaluation of the individual "on" and "off" times. As shown in Fig.7(b), the resulting autocorrelation function is fitted to a single exponential decay, yielding a B/A value of 0.2 and k of 2.17 s^{-1}. Given the excitation rate constant k_{ex} of 2.2x10^4 s^{-1} (38.5 W/cm^2 was used) and the fluorescence decay k_{21} of 3.28x10^8 s^{-1} determined in the excited state lifetime measurement, k_{et} and k_{bet} are evaluated to be 5.4x10^3 and 1.8 s^{-1}, respectively, according to eqs.2,3,7, and 9. The IFET and back ET rate constants with the "on" and "off" times for the examples in Fig.5(b-d) are listed in Table 1. For comparison, the corresponding lifetime measurements are also listed. A more efficient IFET is apparently accompanied by a shorter excited state lifetime.

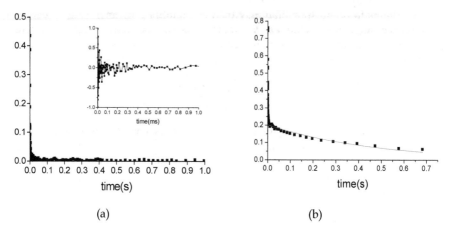

(a)   (b)

Fig. 7. Autocorrelation function of fluorescence intensity from single oxazine 1 molecules (a) on bare coverslip, (b) on TiO$_2$ NPs-coated coverslip. The inset in (a) is the enlarged trace within the range of 1 ms.

	Lifetime/ns	τ_{on} (s)	τ_{off} (s)	k_{et} (s^{-1})	k_{bet} (s^{-1})
A	4.0	-	-	-	-
B	3.4	86.02	1.43	1.6 x10^2	0.7
C	3.1	2.75	0.55	5.4 x10^3	1.8
D	2.9	0.49	0.08	3.2 x10^4	12.0

Table 1. The excited state lifetimes and kinetic data for the single-molecule traces shown in Fig. 6.

As with the above examples, 100 single dye molecules are successively analyzed. The resulting IFET and back ET rate constants are displayed in the form of histogram (Fig.8(a) and (b)), yielding a range of 10^2-10^4 and 0.1-10 s^{-1}, respectively. The distributions are fitted with an individual single-exponential function to yield an average value of (1.0±0.1)x10^4 and 4.7±0.9 s^{-1}, which are the upper limit of the IFET and back ET rate constants among these 100 single molecules analyzed, if the unknown contributions of rotational and translational motion are considered. The obtained average rate of electron transfer is much slower than the fluorescence relaxation. That is why no statistical difference of the fluorescence lifetimes of the dye is found between TiO$_2$ and bare coverslip. The ET rate constant distribution could be affected by different orientation and distance between dye molecule and TiO$_2$ NPs. The weak coupling between electron donor and acceptor may be caused by physisorption between the dye molecule and the TiO_2 NPs or a disfavored energy system for the dye electron jumping into the conduction band of the semiconductor. The resulting ET quantum yield as small as 3.1×10^{-5} is difficult to be detected in the ensemble system. Nevertheless, such slow electron transfer events are detectable at a single molecule level as demonstrated in this work.

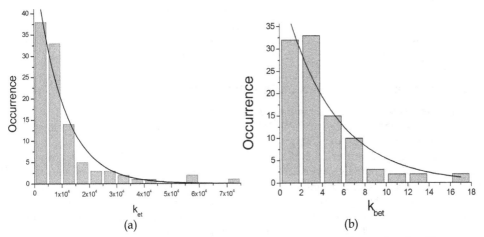

Fig. 8. The histograms of (a) k_{et} and (b) k_{bet} determined among 100 dye molecules. The average values of $(1.0 \pm 0.1) \times 10^4$ and 4.74 s^{-1} are evaluated by a fit to single-exponential function.

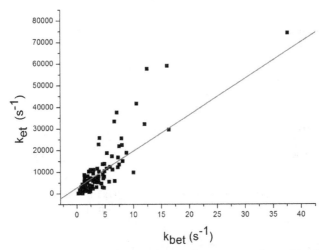

Fig. 9. A linear correlation between photo-induced electron transfer and back electron transfer rate constant.

The process of photo-induced ET involves charge ejection from the oxazine 1 LUMO (~ 2.38 ev) into a large energetically accessible density of states within the conduction band of

TiO$_2$(~4.4 ev), while the back ET involves thermal relaxation of electrons from the conduction band or from a local trap (energetically discrete states) back to the singly occupied molecular orbital (SOMO) of the oxazine 1 cation.[37] It is interesting to find a linear correlation with a slope of 1.7x10^3 between IFET and back ET rate constants, as shown in Fig.9. Despite difference of the mechanisms, k_{bet} increases almost in proportion to k_{et}. Such a strong correlation between forward and backward ET rate constants suggests that for different dye molecules the ET energetics remains the same but the electronic coupling between the excited state of the dye molecules and the conduction band of the solid film varies widely (Cotlet et al., 2004). Both forward and backward ET processes are affected similarly by geometric distance and orientation between electron donor and acceptor.

4. Fluorescence intermittency and electron transfer by quantum dots

4.1 Fluorescence intermittency and lifetime determination

Three different sizes of CdSe/ZnS core/shell QDs were used. Each size was estimated by averaging over 100 individual QDs images obtained by transmission emission spectroscopy (TEM), yielding the diameters of 3.6±0.6, 4.6±0.7, and 6.4±0.8 nm, which are denoted as A, B, and C size, respectively, for convenience. Each kind was then characterized by UV/Vis and fluorescence spectrophotometers to obtain its corresponding absorption and emission spectra. As shown in Fig.10(a) and (b), a smaller size of QDs leads to emission spectrum shifted to shorter wavelength. From their first exciton absorption bands at 500, 544, and 601 nm, the diameter for the CdSe core size was estimated to be 2.4, 2.9, and 4.6 nm (Yu et al., 2003), respectively, sharing about 25-37% of the whole volume. In addition, given the band gaps determined from the absorption bands and the highest occupied molecular orbital (HOMO) potential of -6.12 ~ -6.15 Ev (Tvrdy et al., 2011), the LUMO potentials of QDs may be estimated to be -4.06, -3.86, and -3.67 eV along the order of decreased size.

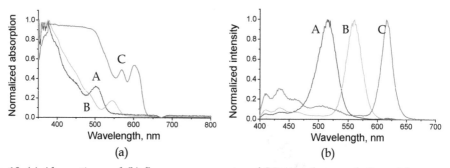

Fig. 10. (a) Absorption and (b) fluorescence spectra of QDs in toluene solution with excitation wavelength fixed at 375 nm. The maximum intensities for both spectra have been normalized. A, B, and C species have the diameters of 3.6, 4.6, and 6.4 nm, respectively.

Each size of QDs was individually spin-coated on bare and TiO$_2$ coverslip. Fig.11 shows an example for the photoluminescence (PL) images within a 24 μm x 24 μm area of the smallest QDs on the glass and TiO$_2$ NPs thin film, as excited at 375 nm. The surface densities of fluorescent QDs on TiO$_2$ were less than those on glass. Their difference becomes more significant with the decreased size of QDs.

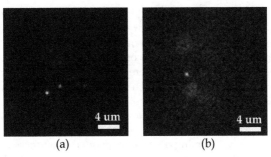

(a)   (b)

Fig. 11. The CCD images of QDs with the diameter of 3.6 nm at 4.5×10^{-11} g/L which was spin-coated on (a) glass and (b) TiO₂ film.

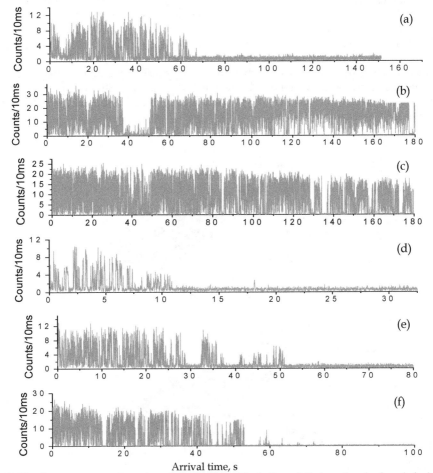

Arrival time, s

Fig. 12. The fluorescence trajectories of single QD with A, B, and C size adsorbed on (a,b,c) glass and (d,e,f) TiO₂ film. The order of increased size is followed from a to c and from d to f.

As a single bright spot was focused, the trajectory of fluorescence intensity was acquired until photobleaching. The trajectory is represented as a number of emitting photons collected within a binning time as a function of the arrival time after the experiment starts. Fig.12 shows the examples for the three sizes of QDs on glass and TiO_2. The bleaching time of the trajectory appears shorter with the decreased size of QDs, showing an average value of 9.4, 19.6, and 34.1 s on TiO_2, which are much shorter than those on glass. In addition, QDs on either surface are characterized by intermittent fluorescence. As compared to those on glass coverslip, QDs on TiO_2 endure shorter on-time (or fluorescing time) events but longer off-time events. This trend is followed along a descending order of size.

The photons collected within a binning time can be plotted as a function of delay time which is defined as the photon arrival time with respect to the excitation pulse. The fluorescence decay for a single QD is thus obtained. Each acquired curve can be applied to a mono-exponential tail-fit, thereby yielding the corresponding lifetime for a selected arrival time slot. For increasing single-to-noise ratio, the on-state lifetime is averaged over the entire trajectory. However, the off-state lifetime cannot be precisely estimated, because its signal is close to the background noise with limited number of photons collected. Fig.13 shows a single QD lifetime determined for different sizes on glass and TiO_2. A smaller size of QDs results in a shorter on-state lifetime on either surface. Given the same size of QDs, the lifetime on TiO_2 appears to be shorter than that on glass. Their lifetime difference increases with the decreased size. As reported previously (Jin et al., 2010a), the trajectories of fluorescence intermittency and lifetime fluctuation are closely correlated. A similar trend is also found in this work.

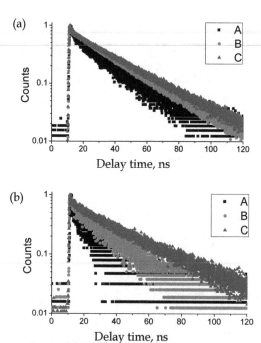

Fig. 13. The fluorescence decay, detected by the TCSPC method, for three types of QDs spin-coated on (a) glass and (b) TiO_2 film. The number of counts is normalized to unity.

Fig.14 shows the lifetime histograms among 20-90 single QDs for the three sizes on glass and TiO$_2$. The corresponding average lifetimes are listed in Table 2. As shown in Fig.14, the smallest QDs on TiO$_2$ have much less on-events than those on glass. For clear lifetime comparison of QDs adsorption between glass and TiO$_2$, each on-event distribution is normalized to unity. The Gaussian-like lifetime histogram has a wide distribution for both glass and TiO$_2$. The lifetime difference for the A type of QDs can be readily differentiated between these two surfaces. As listed in Table 2, their average lifetimes correspond to 19.3 and 14.9 s. In contrast, a tiny lifetime difference between 25.7 and 25.5 s for the C type of QDs is buried in a large uncertainty.

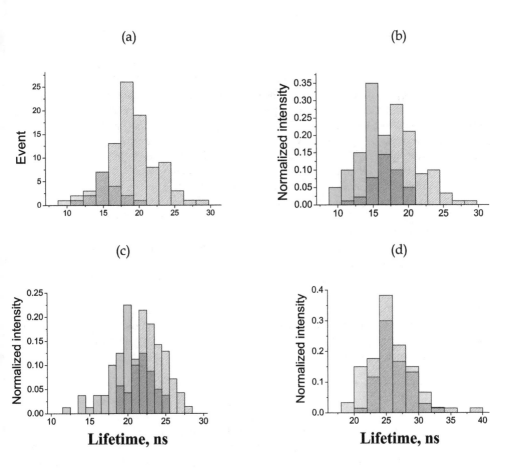

Fig. 14. The distributions of fluorescence lifetime for (a,b) QDs A and (c,d) QDs B and C. (a) comparison of on-event occurrence for QDs A between glass and TiO$_2$. (b,c,d) each area of distribution is normalized to unity. The lifetime distributions of QDs on glass and TiO$_2$ are displayed in red and blue, respectively.

Quantum Dots	Size	Substrate	Amount	Lifetime, ns
A	$3.6 \pm 0.6\,nm$	Glass	90	19.3 ± 3.2
		TiO_2 film	20	14.9 ± 2.4
B	$4.6 \pm 0.7\,nm$	Glass	70	22.6 ± 2.1
		TiO_2 film	80	19.6 ± 2.7
C	6.4 ± 0.8 nm	Glass	68	25.7 ± 2.2
		TiO_2 film	60	25.5 ± 3.2

Table 2. Size-dependence of on-state lifetimes of quantum dots (QDs) on glass and TiO_2 film which are averaged over a quantity of single QDs.

4.2 Interfacial electron transfer

Upon excitation at 375 nm, a QD electron is pumped to the conduction band forming an exciton. The energy gained from recombination of electron and hole will be released radiatively or nonradiatively. However, the excited electron may be feasibly scattered out of its state in the conduction band and be prolonged for recombination. The excited electron probably undergoes resonant tunneling to a trapped state in the shell or nonresonant transition to another trapped state in or outside the QD (Hartmann et al., 2011; Krauss & Peterson, 2010; Jin et al., 2010b; Kuno et al., 2001). The off state of QD is formed, as the charged hole remains. When a second electron-hole pair is generated by a second light pulse or other processes, the energy released from recombination of electron and hole may transfer to the charged hole or trapped electron to cause Auger relaxation. Its relaxation rate is expected to be faster than the PL rate. Given a QD with the core radius of 2 nm, the Auger relaxation rate was estimated to be 100 times larger than the radiative decay rate (Hartmann et al., 2011). The fluorescence fluctuation is obviously affected by the Auger relaxation process that is expected to be <100 ps.

As shown in Fig.12, the on-time events of fluorescence intermittency for QDs on TiO_2 are more significantly suppressed than those on glass. The shortened on events are expected to be caused by the ET from QDs to the TiO_2 film. The analogous phenomena have been reported elsewhere (Hamada et al., 2010; Jin & Lian, 2009). The more rapid the ET is, the shorter the on-state lifetime becomes. The fluorescence lifetime may be estimated by (Jin & Lian, 2009; Kamat, 2008; Robel et al., 2006)

$$\tau = \frac{1}{k_r + k_A + k_{ET}} \qquad (10)$$

where k_r, k_A, and k_{ET} denote intrinsic decay rate of radiation, Auger relaxation rate, and ET rate. When QD is adsorbed on glass, the ET rate is assumed to be zero. The fluorescence fluctuation is dominated by the Auger relaxation. Thus, given the lifetime measurements on both glass and TiO_2 and assumption of the same Auger relaxation rate, the ET rate constant from QDs to TiO_2 can be estimated by the reciprocal of the lifetime difference. The resulting ET rate constants are $(1.5\pm1.4)\times10^7$ and $(6.8\pm8.1)\times10^6$ s^{-1} for the QDs A and B, respectively. A large uncertainty is caused by a wide lifetime distribution. The ET rates depend on the QDs size. The smaller QDs have a twice larger rate constant. However, the ET rate constant for the largest size cannot be determined precisely, due to a slight lifetime difference but with large uncertainty. The ET quantum yield Φ may then be estimated as 22.6 and 13.3% for the A and B sizes, respectively, according to the following equation,

$$\Phi = \frac{k_{ET}}{k_r + k_A + k_{ET}} = k_{ET}\tau \tag{11}$$

The larger QDs result in a smaller quantum yield.

In an analogous experiment, Jin and Lian obtained an average ET rate of 3.2×10^7 s⁻¹ from CdSe/ZnS core/shell QDs with capped carboxylic acid functional groups (Jin & Lian, 2009). Their size was estimated to have core diameter of 4.0 nm based on the first exciton peak at 585 nm. While considering the size dependence, our result is about ten times smaller. It might be caused by the additional carboxylic acid functional groups which can speed up the ET rates.

Different from the method of lifetime measurement, autocorrelation of fluorescence intensities can be alternatively used to quantify the kinetic rate constants in on-off transition, as described in Section III.C (Chen et al., 2010). By analyzing exponential autocorrelation of fluorescence trajectory under a three-level energy system, the forward and back ET rate constants of single oxazine 1 dye/TiO₂ film were reported above (Chen et al., 2010). As the uncertainty was considered, there was no statistical difference of the lifetime measurements for the single dye adsorption between glass and TiO₂. Such a small ET activity can be indeed quantified by analyzing the autocorrelation function. Unfortunately, this method cannot be applied effectively to the QDs case, because the kinetic system involves multiple manifolds that make analysis more complicated.

The non-exponential fluorescence fluctuation was reported in single semiconductor QDs early in 1996 (Nirmal et al., 1996). To explain such fluorescence intermittency, Efros *et al.* (Efros & Rosen, 1997) proposed an Auger ionization model, in which an electron (hole) ejection outside the core QDs is caused by nonradiative relaxation of a bi-exciton. However, Auger ionization process would lead to a single exponential probability distribution of 'on' events, which is against the power-law distributions and the large dynamic range of time scale observed experimentally (Kuno et al., 2000, 2001). Nesbitt and coworkers later investigated the detailed kinetics of fluorescence intermittency in colloidal CdSe QDs and evaluated several related models at the single molecule level. They concluded that the kinetics of electron or hole tunneling to trap sites with environmental fluctuation should be more appropriate to account for the blinking phenomena (Kuno et al., 2001). Frantsuzov and Marcus (Frantsuzov & Marcus, 2005) further suggested a model regarding fluctuation of nonradiative recombination rate to account for the unanswered problem for a continuous distribution of relaxation times.

To compare the blinking activity for a single QD, probability density P(t) is defined to indicate the blinking frequency between the on and off states. The probability density P(t) of a QD at on or off states for duration time t may be calculated by(Kuno et al., 2001; Cui et al., 2008; Jin & Lian, 2009; Jin et al., 2010a)

$$P_i(t) = \frac{N_i(t)}{N_{i,tot}\Delta t_{av}} \tag{12}$$

where i denotes on or off states, N(t) the number of on or off events of duration time t, N_{tot} the total number of on or off events, and Δt_{av} the average time between the nearest neighbor events. The threshold fluorescence intensity to separate the on and off states is set at 3σ. σ is the standard deviation of the background fluorescence intensity which can be fitted with a Gaussian function.

Fig.15 shows a fluorescence trajectory with a threshold intensity and its corresponding blinking frequency for a single QD (3.6 nm) on glass and TiO_2. The subsequent on-state and off-state probability densities accumulated over 10 single QDs for each species are displayed in Fig.16 and Fig.17, which show similar behavior as a single QD but with more data points to reduce uncertainty. The P(t) distribution at the on state for each size under either surface condition essentially follows power law statistics at the short time but deviates downward at the long time tails. The bending tail phenomena are similar to those reported (Tang & Marcus, 2005a, b; Cui et al., 2008; Peterson & Nesbitt, 2009; Jin et al., 2010a). These on-state distributions can be fitted by a truncated power law, as expressed by (Tang & Marcus, 2005a, b; Cui et al., 2008; Peterson & Nesbitt, 2009; Jin et al., 2010a)

$$P_i(t) = Dt^{-m_i} \exp(-\Gamma_i t) \qquad (13)$$

where D is the amplitude associated with electronic coupling and other factors, m_i the power law exponent for the on state, and Γ the saturation rate. The truncated power law was developed by Marcus and coworkers for interpreting the blinking behavior of QD which was attributed to the ET process between a QD and its localized surface states (Tang & Marcus, 2005a, b). According to eq.13, the fitting parameters of m_{on} and Γ_{on} are listed in Table 3. The QDs on TiO_2 apparently result in larger Γ values than those on glass. In addition, the trend is found that a smaller QD may have a larger Γ. As for m_{on}, the obtained range is from 0.70 to 0.93, smaller than 1.5 as expected by Marcus model (Tang & Marcus, 2005a, b; Cui et al., 2008). Such deviation for m_{on} was also found by the Lian group in a similar experiment (Jin et al., 2010a; Jin et al., 2010b). Note that the power law distribution with a bending tail in the long time region is solely found at the on states. In contrast, the off-state probability density may be fit to a simple power law statistics expressed by,

$$P_i(t) = Et^{-m_i} \qquad (14)$$

where E is a scaling coefficient and m_i is the power law exponent for the off state. A similar trend for both on- and off-state distributions was analogously found elsewhere (Cui et al., 2008; Peterson & Nesbitt, 2009). As listed in Table 3, the obtained m_{off} yields a smaller value when QDs are adsorbed on TiO_2. They lie in the range of 1.6-2.1, which are consistent with those reported (Kuno et al., 2001; Cui et al., 2008; Peterson & Nesbitt, 2009).

The on-time saturation rate should be associated with the ET rate. According to Marcus model (Tang & Marcus, 2005a, b), the free energy curves of light emitting state and dark state can be represented by an individual parabola along a reaction coordinate, which is assumed to have the same curvature. Then, Γ_{on} can be related to the free energy change ΔG_{ET} based on the ET process. That is (Tang & Marcus, 2005a, b; Cui et al., 2008),

$$\Gamma_{on} = \frac{(\lambda + \Delta G_{ET})^2}{8t_{diff} \lambda k_B T} \qquad (15)$$

where λ is the system reorganization energy, t_{diff} the diffusion correlation time constant for motion on a parabolic energy surface, k_B the Boltzmann constant, and T the absolute temperature. Given the conduction band of -4.41 eV for TiO_2 NPs and the LUMO potentials of QDs, -3.67 and -3.86 eV for the A and B sizes, respectively, the corresponding $-\Delta G_{ET}$ may

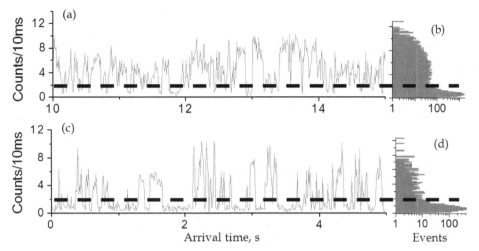

Fig. 15. The fluorescence trajectory and corresponding on/off blinking frequency distribution of single QD A on (a,b) glass and (c,d) TiO₂ film. The black lines denote the intensity thresholds to separate the on and off state which are set at a level 3σ above the background noise. σ is the standard deviation of background noise.

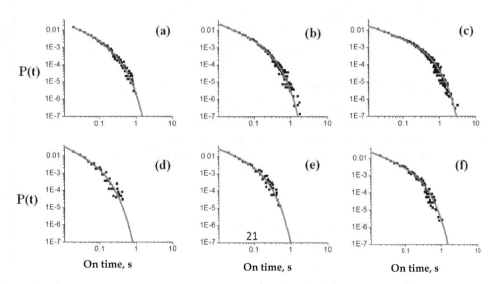

Fig. 16. The on-state probability density of 10 single QDs with A, B, and C size on (a,b,c) glass and (d,e,f) TiO₂ film. The order of increased size is followed from a to c and from d to f. The spots denote experimental data and lines denote simulation by truncated power law distribution.

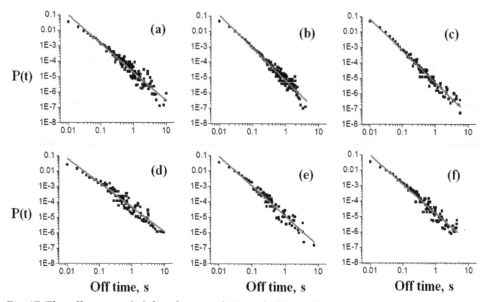

Fig. 17. The off-state probability density of 10 single QDs with A, B, and C size on (a,b,c) glass and (d,e,f) TiO₂ film. The order of increased size is followed from a to c and from d to f. The spots denote experimental data and lines denote simulation by power law distribution.

Quantum dots	Substrate	m_{on}	m_{off}	$1/\Gamma_{on}$,ms
A	Glass	0.910 ± 0.005	1.85 ± 0.04	181.9 ± 3.1
	TiO₂ film	0.925 ± 0.006	1.58 ± 0.05	75.7 ± 1.3
B	Glass	0.803 ± 0.002	2.07 ± 0.05	188.6 ± 1.9
	TiO₂ film	0.785 ± 0.005	1.91 ± 0.06	108.4 ± 2.1
C	Glass	0.699 ± 0.002	2.04 ± 0.04	355.4 ± 4.3
	TiO₂ film	0.760 ± 0.004	1.87 ± 0.05	172.0 ± 3.1

Table 3. The fitting parameters of 10 single quantum dots at the on state in terms of truncated power law distribution and off state in terms of power law distribution.

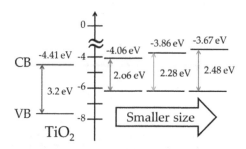

Fig. 18. The energy diagram of TiO₂ and QDs with A, B, and C size.

be estimated to be 0.74 and 0.55 eV. The related energy diagram is displayed in Fig.18. For a smaller QD, the larger conduction band gap between QD and TiO₂ can induce a larger driving force to facilitate the ET process (Tvrdy et al., 2011). If λ and t_{diff} remain constant, substituting ΔG_{ET} and Γ_{on} into eq.15 for different size of QDs yields λ to be 636 and -416 meV, of which only the positive value is meaningful.

4.3 Model prediction of electron transfer
In the following is the Marcus model which has been successfully used to describe the ET kinetics for the systems of organic dyes coupled to various metal oxides (She et al., 2005; Tvrdy et al., 2011),

$$k_{ET} = \frac{4\pi^2}{h} \int_{-\infty}^{\infty} \rho(E) |H(E)|^2 \frac{1}{\sqrt{4\pi\lambda k_B T}} \exp(\frac{(1+\Delta G + E)^2}{4\lambda k_B T}) dE \qquad (16)$$

where H(E) is the overlap matrix element, $\rho(E)$ the density of electron accepting states, h the Planck's constant, and ΔG the free energy change of the system, which is composed of three factors. They are (1) the energy change between initial and final electronic states, equivalent to ΔG_{ET} mentioned in eq.15, (2) the free energy difference between nonneutral donating and accepting species in the ET process, and (3) the free energy of coulombic interaction for electron and hole separation (Tvrdy et al., 2011). Among them, only ΔG_{ET} can be measured experimentally. Because of similarity as the work by the Kamat group (Tvrdy et al., 2011), the contributions of the second and third factors are referred to their work. That is,

$$\Delta G = \Delta G_{ET} + \frac{e^2}{2R_{QD}} + 2.2 \frac{e^2}{\varepsilon_{QD} R_{QD}} - \frac{e^2}{4(R_{QD}+s)} \frac{\varepsilon_{TiO_2}-1}{\varepsilon_{TiO_2}+1} \qquad (17)$$

where e is the elementary charge, R_{QD} and ε_{QD} are the radius and dielectric permittivity of the QD, ε_{TiO2} is the dielectric permittivity of TiO₂, and s is the separation distance between QD and TiO₂. Given s, assumed to be the same as reported (Tvrdy et al., 2011), and the data of ΔG_{ET}, R_{QD}, ε_{QD} and ε_{TiO2}, ΔG is estimated to be -0.22, -0.143, and -0.054 eV for the A, B, and C sizes of QDs, respectively. As compared to ΔG_{ET}, the driving force for moving electron from QD to TiO₂ is suppressed after taking into account the additional contributions in eq.17.

In a perfect semiconductor crystal, the density of unoccupied states $\rho(E)$ is given as (She et al., 2005; Tvrdy et al., 2011)

$$\rho(E) = V_o \frac{(2m_e^*)^{3/2}}{2\pi\hbar^3} \sqrt{E} \qquad (18)$$

where V_o is the effective volume, known to be 34.9 Å³ for TiO₂ crystal, m_e^* the electron effective mass, equivalent to 10 m_0 (m_0 is the mass of free electron) (She et al., 2005), and ℏ is $h/2\pi$. For a TiO₂ nanoparticle with high surface to volume ratio, the density of states in eq.18 requires modification by considering the defect states which are treated as a Gaussian distribution of width Δ. (She et al., 2005; Tvrdy et al., 2011) The modified density of states $\rho_D(E)$ is then expressed as

$$\rho_D(E) = \int_0^{\infty} \rho(E') \frac{1}{\Delta\sqrt{2\pi}} \exp(\frac{(E-E')^2}{2\Delta^2}) dE' \qquad (19)$$

Given a constant H(E), substituting eqs.18 and 19 into eq.16 yields an explicit relation between k_{ET} and ΔG.

As reported (She et al., 2005), when the dye/metal oxide system was surrounded by a buffer layer, the reorganization energy λ increases to 100-500 meV, because additional energy is required for system rearrangement. In this work, CdSe/ZnS QDs are spin-coated on the TiO$_2$ NPs film which is exposed to the air. The requirement of reorganization energy should be small. Therefore, the λ value of 636 meV in the estimate (eq.15) seems to be unreasonable. When ΔG_{ET} is replaced by ΔG, λ is obtained to be 178 and -248 meV based on eq.15. The selected λ of 178 meV is more acceptable than the one obtained with ΔG_{ET} substituted. The width Δ of defect states for the TiO$_2$ NPs is insensitive in the k_{ET} calculation by eq.16. We adopt the same Δ of 50 meV as reported (Tvrdy et al., 2011). Given H(E) assumed to have 0.83 cm^{-1} and the data of ΔG, λ, and Δ, k_{ET} is optimized to be 1.42x10^7, 6.80x10^6, and 1.86x10^6 s^{-1} for three increased sizes of QDs. The first two results agree very well with the experimental findings. The ET rate constant for the C type of QDs cannot be precisely determined experimentally, but may be estimated with the aid of model prediction.

For our system, CdSe/ZnS core/shell QDs are spin-coated on the TiO$_2$ thin film. Unlike this preparation procedure, Kamat and coworkers immersed TiO$_2$ film in the colloidal CdSe QDs solution to make a tight contact between donor and acceptor and then measured the electron transfer rates under a vacuum condition (Tvrdy et al., 2011). Therefore, the overlap matrix elements, H(E), which is associated with the coupling between electron donating and accepting states, must make difference. In our work, the core CdSe QD and TiO$_2$ have a loose contact and thus a smaller H(E) of 0.83 cm^{-1} is obtained. In contrast, a much larger value of 57 cm^{-1} was adopted by the Kamat group (Tvrdy et al., 2011). That is why the ET rates obtained herein are relatively slower by a factor of 10^4.

5. Concluding remarks

This chapter describes IFET induced by a single dye molecule or a single QD which is individually adsorbed on the TiO$_2$ NPs film. The fluorescence lifetimes determined among different single oxazine 1 dye molecules are widely spread, because of micro-environmental influence. These lifetimes are in proximity to those measured on the bare coverslip, indicative of the IFET inefficiency for those dye molecules sampled in this work. However, some molecules may proceed via very efficient IFET process, but fail to be detected. Due to a shorter triplet excursion, oxazine 1-TiO$_2$ NPs system is treated effectively as a three-level system upon irradiation. The exponential autocorrelation function may thus be analyzed to quantify the related kinetic rate constants in an on-off transition. The IFET processes are found to be inhomogeneous, with a rate constant varying from molecule to molecule. The reactivity and rate of ET fluctuation of the same single molecule are the main source to result in fluorescence intensity fluctuation. These phenomena, which are obscured in the ensemble-averaged system, are attributed to micro-environment variation for each single molecule. The oxazine 1 dye is apparently unsuitable for application to the DSSC design, because of its lower ET rates. Nevertheless, the single molecule spectroscopy provides a potential tool looking into the microscopic ET behaviors for different dye molecules to facilitate the working efficiency for the cell design. In addition, it is capable of detecting a low ET quantum yield, which is difficult to measure with conventional ensemble-averaged methods.

The second part of this chapter describes the ET from QDs to TiO_2 NPs film. The ET kinetics depends on the size of CdSe/ZnS QDs. The trajectories of fluorescence intermittency of three different sizes of QDs on glass and TiO_2 are acquired and the subsequent fluorescence lifetimes are determined. While assuming the lack of electron transfer for the QD on glass, the ET rates from QD to TiO_2 may be inferred in terms of reciprocal of the lifetime difference. The following trend is found: the smaller the size of QDs, the larger the ET rate constants. The distribution of off-time probability density versus the arrival time is fit to a simple power law statistics. However, the plot of on-time probability density can be characterized by a truncated power law distribution. Marcus's electron transfer model is employed to fit the bending tail behavior and to further calculate the ET rate constants, which show consistency with our experimental findings.

6. Acknowledgments

This work is supported by National Science Council, Taiwan, Republic of China under contract no. NSC 99-2113-M-001-025-MY3 and National Taiwan University, Ministry of Education.

7. References

Ambrose, W. P., P. M. Goodwin, J. C. Martin & R. A. Keller (1994) Single-molecule Detection and Photochemistry on a Surface Using Near-field Optical excitation. *Physical Review Letters*, 72, 1, (Jan 1994), 160-163, 0031-9007

Arden, W. & P. Fromherz (1980) Photosensitization of Semiconductor Electrode by Cyanine Dye in Lipid Bilayer. *Journal of the Electrochemical Society*, 127, 2, 1980), 370-378, 0013-4651

Baker, D. R. & P. V. Kamat (2009) Photosensitization of TiO2 Nanostructures with CdS Quantum Dots: Particulate versus Tubular Support Architectures. *Advanced Functional Materials*, 19, 5, (Mar 2009), 805-811, 1616-301X

Bell, T. D. M., C. Pagba, M. Myahkostupov, J. Hofkens & P. Piotrowiak (2006) Inhomogeneity of electron injection rates in dye-sensitized TiO2: Comparison of the mesoporous film and single nanoparticle Behavior. *Journal of Physical Chemistry B*, 110, 50, (Dec 2006), 25314-25321, 1520-6106

Biesmans, G., M. Vanderauweraer, C. Cathry & F. C. Deschryver (1992) On the Photosensitized Injection of Electrons into SNO2 from Cyanine Dyes Incorporated in Langmuir-Blodgett-Films. *Chemical Physics*, 160, 1, (Feb 1992), 97-121, 0301-0104

Biesmans, G., M. Vanderauweraer, C. Cathry, D. Meerschaut, F. C. Deschryver, W. Storck & F. Willig (1991) Photosensitized Electron Injection from Xanthene Dyes Incorporated in Langmuir-Blodgett-Films into SNO2 Electrodes. *Journal of Physical Chemistry*, 95, 9, (May 1991), 3771-3779, 0022-3654

Bisquert, J., A. Zaban & P. Salvador (2002) Analysis of the mechanisms of electron recombination in nanoporous TiO2 dye-sensitized solar cells. Nonequilibrium steady-state statistics and interfacial electron transfer via surface states. *Journal of Physical Chemistry B*, 106, 34, (Aug 2002), 8774-8782, 1520-6106

Cahen, D., G. Hodes, M. Gratzel, J. F. Guillemoles & I. Riess (2000) Nature of photovoltaic action in dye-sensitized solar cells. *Journal of Physical Chemistry B*, 104, 9, (Mar 2000), 2053-2059, 1089-5647

Chen, Y. J., H. Y. Tzeng, H. F. Fan, M. S. Chen, J. S. Huang & K. C. Lin (2010) Photoinduced Electron Transfer of Oxazine 1/TiO2 Nanoparticles at Single Molecule Level by Using Confocal Fluorescence Microscopy. *Langmuir*, 26, 11, (Jun 2010), 9050-9060, 0743-7463

Choi, J. J., Y. F. Lim, M. B. Santiago-Berrios, M. Oh, B. R. Hyun, L. F. Sung, A. C. Bartnik, A. Goedhart, G. G. Malliaras, H. D. Abruna, F. W. Wise & T. Hanrath (2009) PbSe Nanocrystal Excitonic Solar Cells. *Nano Letters*, 9, 11, (Nov 2009), 3749-3755, 1530-6984

Cotlet, M., S. Masuo, G. B. Luo, J. Hofkens, M. Van der Auweraer, J. Verhoeven, K. Mullen, X. L. S. Xie & F. De Schryver (2004) Probing conformational dynamics in single donor-acceptor synthetic molecules by means of photoinduced reversible electron transfer. *Proceedings of the National Academy of Sciences of the United States of America*, 101, 40, (Oct 2004), 14343-14348, 0027-8424

Cotlet, M., T. Vosch, S. Habuchi, T. Weil, K. Mullen, J. Hofkens & F. De Schryver (2005) Probing intramolecular Forster resonance energy transfer in a naphthaleneimide-peryleneimide-terrylenediimide-based dendrimer by ensemble and single-molecule fluorescence spectroscopy. *Journal of the American Chemical Society*, 127, 27, (Jul 2005), 9760-9768, 0002-7863

Cui, S. C., T. Tachikawa, M. Fujitsuka & T. Majima (2008) Interfacial Electron Transfer Dynamics in a Single CdTe Quantum Dot-Pyromellitimide Conjugate. *Journal of Physical Chemistry C*, 112, 49, (Dec 2008), 19625-19634, 1932-7447

Efros, A. L. & M. Rosen (1997) Random telegraph signal in the photoluminescence intensity of a single quantum dot. *Physical Review Letters*, 78, 6, (Feb 1997), 1110-1113, 0031-9007

Fan, S. Q., B. Fang, J. H. Kim, J. J. Kim, J. S. Yu & J. Ko (2010) Hierarchical nanostructured spherical carbon with hollow core/mesoporous shell as a highly efficient counter electrode in CdSe quantum-dot-sensitized solar cells. *Applied Physics Letters*, 96, 6, (Feb 2010), 0003-6951

Ferrere, S. & B. A. Gregg (2001) Large increases in photocurrents and solar conversion efficiencies by UV illumination of dye sensitized solar cells. *Journal of Physical Chemistry B*, 105, 32, (Aug 2001), 7602-7605, 1089-5647

Flors, C., I. Oesterling, T. Schnitzler, E. Fron, G. Schweitzer, M. Sliwa, A. Herrmann, M. van der Auweraer, F. C. de Schryver, K. Mullen & J. Hofkens (2007) Energy and electron transfer in ethynylene bridged perylene diimide multichromophores. *Journal of Physical Chemistry C*, 111, 12, (Mar 2007), 4861-4870, 1932-7447

Frantsuzov, P. A. & R. A. Marcus (2005) Explanation of quantum dot blinking without the long-lived trap hypothesis. *Physical Review B*, 72, 15, (Oct 2005), 1098-0121

Gaiduk, A., R. Kuhnemuth, S. Felekyan, M. Antonik, W. Becker, V. Kudryavtsev, C. Sandhagen & C. A. M. Seidel (2007) Fluorescence detection with high time resolution: From optical microscopy to simultaneous force and fluorescence spectroscopy. *Microscopy Research and Technique,* 70, 5, (May 2007), 433-441, 1059-910X

Garcia-Parajo, M. F., G. M. J. Segers-Nolten, J. A. Veerman, J. Greve & N. F. van Hulst (2000) Real-time light-driven dynamics of the fluorescence emission in single green fluorescent protein molecules. *Proceedings of the National Academy of Sciences of the United States of America,* 97, 13, (Jun 2000), 7237-7242, 0027-8424

Gratzel, M. (2001) Photoelectrochemical cells. *Nature,* 414, 6861, (Nov 2001), 338-344, 0028-0836

Gratzel, M. (2003) Dye-sensitized solar cells. *Journal of Photochemistry and Photobiology C-Photochemistry Reviews,* 4, 2, (Oct 2003), 145-153, 1389-5567

Gratzel, M. (2005) Mesoscopic solar cells for electricity and hydrogen production from sunlight. *Chemistry Letters,* 34, 1, (Jan 2005), 8-13, 0366-7022

Haase, M., C. G. Hubner, E. Reuther, A. Herrmann, K. Mullen & T. Basche (2004) Exponential and power-law kinetics in single-molecule fluorescence intermittency. *Journal of Physical Chemistry B,* 108, 29, (Jul 2004), 10445-10450, 1520-6106

Hagfeldt, A. & M. Gratzel (2000) Molecular photovoltaics. *Accounts of Chemical Research,* 33, 5, (May 2000), 269-277, 0001-4842

Hamada, M., S. Nakanishi, T. Itoh, M. Ishikawa & V. Biju (2010) Blinking Suppression in CdSe/ZnS Single Quantum Dots by TiO2 Nanoparticles. *Acs Nano,* 4, 8, (Aug 2010), 4445-4454, 1936-0851

Hara, K., H. Horiuchi, R. Katoh, L. P. Singh, H. Sugihara, K. Sayama, S. Murata, M. Tachiya & H. Arakawa (2002) Effect of the ligand structure on the efficiency of electron injection from excited Ru-phenanthroline complexes to nanocrystalline TiO2 films. *Journal of Physical Chemistry B,* 106, 2, (Jan 2002), 374-379, 1520-6106

Hartmann, T., V. I. Yudson & P. Reineker (2011) Model for the off-time distribution in blinking quantum dots. *Journal of Luminescence,* 131, 3, 2011), 379-381, 0022-2313

Holman, M. W. & D. M. Adams (2004) Using single-molecule fluorescence spectroscopy to study electron transfer. *ChemPhysChem,* 5, 12, (Dec 2004), 1831-1836, 1439-4235

Jin, S. Y., J. C. Hsiang, H. M. Zhu, N. H. Song, R. M. Dickson & T. Q. Lian (2010a) Correlated single quantum dot blinking and interfacial electron transfer dynamics. *Chemical Science,* 1, 4, (Oct 2010a), 519-526, 2041-6520

Jin, S. Y. & T. Q. Lian (2009) Electron Transfer Dynamics from Single CdSe/ZnS Quantum Dots to TiO2 Nanoparticles. *Nano Letters,* 9, 6, (Jun 2009), 2448-2454, 1530-6984

Jin, S. Y., N. H. Song & T. Q. Lian (2010b) Suppressed Blinking Dynamics of Single QDs on ITO. *Acs Nano,* 4, 3, (Mar 2010b), 1545-1552, 1936-0851

Ju, T., R. L. Graham, G. M. Zhai, Y. W. Rodriguez, A. J. Breeze, L. L. Yang, G. B. Alers & S. A. Carter (2010) High efficiency mesoporous titanium oxide PbS quantum dot solar cells at low temperature. *Applied Physics Letters,* 97, 4, (Jul 2010), 0003-6951

Kamat, P. V. (2008) Quantum Dot Solar Cells. Semiconductor Nanocrystals as Light Harvesters. *Journal of Physical Chemistry C,* 112, 48, (Dec 2008), 18737-18753, 1932-7447

Kim, S. J., W. J. Kim, Y. Sahoo, A. N. Cartwright & P. N. Prasad (2008) Multiple exciton generation and electrical extraction from a PbSe quantum dot photoconductor. *Applied Physics Letters,* 92, 3, (Jan 2008), 0003-6951

Kohn, F., J. Hofkens, R. Gronheid, M. Van der Auweraer & F. C. De Schryver (2002) Parameters influencing the on- and off-times in the fluorescence intensity traces of single cyanine dye molecules. *Journal of Physical Chemistry A*, 106, 19, (May 2002), 4808-4814, 1089-5639

Krauss, T. D. & J. J. Peterson (2010) Bright Future for Fluorescence Blinking in Semiconductor Nanocrystals. *Journal of Physical Chemistry Letters*, 1, 9, (May 2010), 1377-1382, 1948-7185

Kulzer, F., S. Kummer, R. Matzke, C. Brauchle & T. Basche (1997) Single-molecule optical switching of terrylene in p-terphenyl. *Nature*, 387, 6634, (Jun 1997), 688-691, 0028-0836

Kuno, M., D. P. Fromm, H. F. Hamann, A. Gallagher & D. J. Nesbitt (2000) Nonexponential "blinking" kinetics of single CdSe quantum dots: A universal power law behavior. *Journal of Chemical Physics*, 112, 7, (Feb 2000), 3117-3120, 0021-9606

Kuno, M., D. P. Fromm, H. F. Hamann, A. Gallagher & D. J. Nesbitt (2001) "On"/"off" fluorescence intermittency of single semiconductor quantum dots. *Journal of Chemical Physics*, 115, 2, (Jul 2001), 1028-1040, 0021-9606

Lee, Y. L. & Y. S. Lo (2009) Highly Efficient Quantum-Dot-Sensitized Solar Cell Based on Co-Sensitization of CdS/CdSe. *Advanced Functional Materials*, 19, 4, (Feb 2009), 604-609, 1616-301X

Luther, J. M., M. C. Beard, Q. Song, M. Law, R. J. Ellingson & A. J. Nozik (2007) Multiple exciton generation in films of electronically coupled PbSe quantum dots. *Nano Letters*, 7, 6, (Jun 2007), 1779-1784, 1530-6984

Luther, J. M., M. Law, M. C. Beard, Q. Song, M. O. Reese, R. J. Ellingson & A. J. Nozik (2008) Schottky Solar Cells Based on Colloidal Nanocrystal Films. *Nano Letters*, 8, 10, (Oct 2008), 3488-3492, 1530-6984

Michalet, X., S. Weiss & M. Jager (2006) Single-molecule fluorescence studies of protein folding and conformational dynamics. *Chemical Reviews*, 106, 5, (May 2006), 1785-1813, 0009-2665

Moerner, W. E. & D. P. Fromm (2003) Methods of single-molecule fluorescence spectroscopy and microscopy. *Review of Scientific Instruments*, 74, 8, (Aug 2003), 3597-3619, 0034-6748

Nirmal, M., B. O. Dabbousi, M. G. Bawendi, J. J. Macklin, J. K. Trautman, T. D. Harris & L. E. Brus (1996) Fluorescence intermittency in single cadmium selenide nanocrystals. *Nature*, 383, 6603, (Oct 1996), 802-804, 0028-0836

Oregan, B. & M. Gratzel (1991) A Low-cost, High-efficiency Solar-cell Based on Dye-sensitized Colloidal TiO_2 Films. *Nature*, 353, 6346, (Oct 1991), 737-740, 0028-0836

Peterson, J. J. & D. J. Nesbitt (2009) Modified Power Law Behavior in Quantum Dot Blinking: A Novel Role for Biexcitons and Auger Ionization. *Nano Letters*, 9, 1, (Jan 2009), 338-345, 1530-6984

Plass, R., S. Pelet, J. Krueger, M. Gratzel & U. Bach (2002) Quantum dot sensitization of organic-inorganic hybrid solar cells. *Journal of Physical Chemistry B*, 106, 31, (Aug 2002), 7578-7580, 1520-6106

Ramakrishna, G., D. A. Jose, D. K. Kumar, A. Das, D. K. Palit & H. N. Ghosh (2005) Strongly coupled ruthenium-polypyridyl complexes for efficient electron injection in dye-sensitized semiconductor nanoparticles. *Journal of Physical Chemistry B*, 109, 32, (Aug 2005), 15445-15453, 1520-6106

Robel, I., V. Subramanian, M. Kuno & P. V. Kamat (2006) Quantum dot solar cells. Harvesting light energy with CdSe nanocrystals molecularly linked to mesoscopic TiO2 films. *Journal of the American Chemical Society*, 128, 7, (Feb 2006), 2385-2393, 0002-7863

Sambur, J. B., T. Novet & B. A. Parkinson (2010) Multiple Exciton Collection in a Sensitized Photovoltaic System. *Science*, 330, 6000, (Oct 2010), 63-66, 0036-8075

Sens, R. & K. H. Drexhage (1981) Fluorescence Quantum Yield of Oxazine and Carbazine Laser-dyes. *Journal of Luminescence*, 24-5, NOV, 1981), 709-712, 0022-2313

She, C. X., N. A. Anderson, J. C. Guo, F. Liu, W. H. Goh, D. T. Chen, D. L. Mohler, Z. Q. Tian, J. T. Hupp & T. Q. Lian (2005) pH-dependent electron transfer from re-bipyridyl complexes to metal oxide nanocrystalline thin films. *Journal of Physical Chemistry B*, 109, 41, (Oct 2005), 19345-19355, 1520-6106

Tang, J. & R. A. Marcus (2005a) Diffusion-controlled electron transfer processes and power-law statistics of fluorescence intermittency of nanoparticles. *Physical Review Letters*, 95, 10, (Sep 2005a), 0031-9007

Tang, J. & R. A. Marcus (2005b) Mechanisms of fluorescence blinking in semiconductor nanocrystal quantum dots. *Journal of Chemical Physics*, 123, 5, (Aug 2005b), 0021-9606

Tvrdy, K., P. A. Frantsuzov & P. V. Kamat (2011) Photoinduced electron transfer from semiconductor quantum dots to metal oxide nanoparticles. *Proceedings of the National Academy of Sciences of the United States of America*, 108, 1, (Jan 2011), 29-34, 0027-8424

VandenBout, D. A., W. T. Yip, D. H. Hu, D. K. Fu, T. M. Swager & P. F. Barbara (1997) Discrete intensity jumps and intramolecular electronic energy transfer in the spectroscopy of single conjugated polymer molecules. *Science*, 277, 5329, (Aug 1997), 1074-1077, 0036-8075

Veerman, J. A., M. F. Garcia-Parajo, L. Kuipers & N. F. van Hulst (1999) Time-varying triplet state lifetimes of single molecules. *Physical Review Letters*, 83, 11, (Sep 1999), 2155-2158, 0031-9007

Wang, Y. M., X. F. Wang, S. K. Ghosh & H. P. Lu (2009) Probing Single-Molecule Interfacial Electron Transfer Dynamics of Porphyrin on TiO2 Nanoparticles. *Journal of the American Chemical Society*, 131, 4, (Feb 2009), 1479-1487, 0002-7863

Wilkinson, F., G. P. Kelly, L. F. V. Ferreira, V. Freire & M. I. Ferreira (1991) Benzophenone Sensitization of Triplet Oxazine and of Delayed Fluorescence by Oxazine in Acetonitrile Solution. *Journal of the Chemical Society-Faraday Transactions*, 87, 4, (Feb 1991), 547-552, 0956-5000

Xie, X. S. & R. C. Dunn (1994) Probing Single-molecule Dynamics. *Science*, 265, 5170, (Jul 1994), 361-364, 0036-8075

Yip, W. T., D. H. Hu, J. Yu, D. A. Vanden Bout & P. F. Barbara (1998) Classifying the photophysical dynamics of single- and multiple-chromophoric molecules by single molecule spectroscopy. *Journal of Physical Chemistry A*, 102, 39, (Sep 1998), 7564-7575, 1089-5639

Yu, P. R., K. Zhu, A. G. Norman, S. Ferrere, A. J. Frank & A. J. Nozik (2006) Nanocrystalline
     TiO2 solar cells sensitized with InAs quantum dots. *Journal of Physical Chemistry B*,
     110, 50, (Dec 2006), 25451-25454, 1520-6106
Yu, W. W., L. H. Qu, W. Z. Guo & X. G. Peng (2003) Experimental determination of the
     extinction coefficient of CdTe, CdSe, and CdS nanocrystals. *Chemistry of Materials*,
     15, 14, (Jul 2003), 2854-2860, 0897-4756

4

Ordered Semiconductor Photoanode Films for Dye-Sensitized Solar Cells Based on Zinc Oxide-Titanium Oxide Hybrid Nanostructures

Xiang-Dong Gao, Cai-Lu Wang, Xiao-Yan Gan and Xiao-Min Li
State Key Lab of High Performance Ceramics and Superfine Microstructures, Shanghai Institute of Ceramics, Chinese Academy of Sciences, Shanghai, P. R. China

1. Introduction

Dye-sensitized solar cell (DSC) is a new type solar cell based on the photoelectric conversion occurred at the organic dye-semiconductor nanoparticle interface. DSC has attracted great research interests due to its high efficiency for energy conversion (11%) and low production cost compared with the traditional Si based photovoltaic cell (O'Regan et al., 1991; Grätzel et al., 2001). In confront of the huge difficulties in propelling the conversion efficiency of DSC up to 15% or higher, very recently, significant research efforts have been redirected to the optimization of photoanode, while others are stick to seeking better sensitizer or solid electrolytes (Mor et al., 2006a; Grätzel et al., 2006; Li et al., 2006; Zhang et al., 2011).

The photoanode is the kernel component of DSC assuming two major functions, supporting dye molecules, and transporting photo-induced electrons to the bottom electrode. Traditionally, the photoanode is the nanoporous thick film (15-20 um in thickness) consisted of TiO_2 nanoparticles, characterized by the disordered nature (Figure 1a). The electron transport in this disordered network of TiO_2 film is rather difficult due to two factors: (1) The presence of a large quantity of electron traps at the nanostructured TiO_2/electrolyte interface, such as the intrinsic defect sites (e.g. oxygen defects or surface states) and the grain boundaries, which may influence the interfacial charge-transfer kinetics. (2) The tortuosity of the electron path in the photoanode formed via the calcinations at high temperature, which may hinder the free transport of electron and increase the electron-hole recombination rate. In contrast, the ideal photoanode film (Figure 1b) should be built up by very thin (10-20 nm) and long (10-20 um) semiconductor nanowire array, possessing no surface states, which should exhibit the total photoelectric conversion efficiency of 24% based on the widely used black dye (N719) and I^2/I^{3-} electrolyte system. However, limited by the current fabrication technology level, it is extremely difficult to realize this photoanode film in a short term. Therefore, the development of some practical measures to optimize the photoanode is fundamentally important. The photoanode film can be optimized further from two aspects, to enhance electron gathering and transporting efficiency and to inhibit the charge recombination at the same time. Figure 1c illustrates the schematic of the hybrid photoanode structure, using 1D ZnO nanowire as the major path for the electron transportation, and using 0D TiO_2 nanoparticles as the supporting framework of dye molecules. By using two kinds of nanostructures assuming two different functions of the photoanode, the efficiency of the hybrid photoanode is expected to improve compared with the traditional nanoparticle-based photoanode.

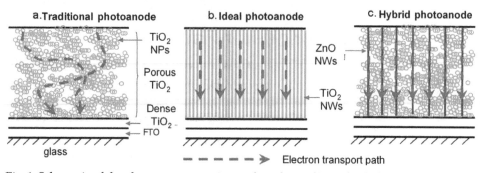

Fig. 1. Schematic of the electron transporting path in the traditional, ideal, and hybrid photoanode. NPs: Nanoparticles; NWs: Nanowires.

Many efforts have been devoted to construct photoanodes of DSCs using ordered semiconductor nanostructures such as nanowires, nanotubes or hierarchical structures. In 2005, M. Law et al. first demonstrated the photoanode film built by a dense array of oriented and crystalline ZnO nanowires (Law et al., 2005). The direct electrical pathways provided by the nanowires ensured the rapid collection of carriers generated throughout the device, and a full Sun efficiency of 1.5% was demonstrated, limited primarily by the surface area of the nanowire array. G. K. Mor et al. reported highly ordered transparent TiO_2 nanotube arrays (6-nm pore diameter, 17-nm wall thickness, and 360-nm length) as the photoanode, which were grown perpendicular to a fluorine-doped tin oxide-coated glass substrate by anodic oxidation of a titanium thin film (Mor et al., 2006b). Although the electrode was only 360-nm-thick, the generated photocurrent was 7.87 mA/cm^2, with a photocurrent efficiency of 2.9%, indicating the great potential of the ordered photoanode. K. Zhu et al. investigated the dynamics of electron transport and recombination in DSCs incorporating oriented TiO_2 nanotube arrays from electrochemically anodized Ti foils, and proved that the nanotube-based DSCs had significantly higher charge-collection efficiencies than the nanoparticle-based counterparts (Zhu et al., 2007). S H Kang et al. developed chemically synthesized TiO_2 nanorod based photoanodes (Kang et al., 2008). These nanorod-based DSCs showed improved photovoltaic properties (6.2 % versus 4.3 % for nanoparticle-based DSCs) owing to the characteristics of slightly enhanced electron transport and predominantly degraded charge recombination. W. Zhang et al. used mesoporous TiO_2 nanofibers with a high surface area of 112 m^2/g prepared by electrospinning technique as the photoanode, and demonstrated the conversion efficiency of 1.82% in solid-state DSCs (Zhang et al., 2010). Recently, C. S. Rustomji et al. designed and constructed photoanodes based on new 3-D configurations of TiO_2 nanotubes, which, unlike prior nanotube-based cells where tubes were grown vertically in a 2-D array, consisted of tubes that extended radially in a 3-D array from a grid of fine titania wires (Rustomji et al., 2010). Its overall efficiency reached 5.0%, and the incident photon-to-current efficiency exceeds 60% over a broad part of the visible spectrum. S. Guldin et al. presented a material assembly route toward double-layer photoanode of DSCs, by coupling a high-surface mesoporous underlayer with an optically and electrically active three-dimensionally periodic TiO_2 photonic crystal overlayer (Guldin et al., 2010). In contrast to earlier studies, the double layer structure exhibited porosity at the mesoporous and the microporous length scales as well as pore and electronic connectivity at

Ordered Semiconductor Photoanode Films for Dye-Sensitized Solar Cells Based on Zinc Oxide-Titanium
Oxide Hybrid Nanostructures

85

all levels. This construct enabled effective dye sensitization, electrolyte infiltration and charge collection from both the mesoporous and the photonic crystal layers.

Apart from various ordered TiO_2 nanostructures, ZnO has also been widely investigated as an alternative of TiO_2 to construct the photoanode of DSCs. For example, E. Hosono et al. pyrolyzed upright-standing sheets of layered hydroxide zinc carbonate, and obtained upstanding nanocrystalline ZnO films with the c-axis parallel to the substrate (Hosono et al., 2005). The corresponding DSCs possessed the conversion efficiency of 3.9%, much higher than the regular efficiency level of ~2% at that time (Rensmo et al., 1997). C. K. Xu et al. reported the photoanode using ultralong ZnO nanowires with up to 33 μm thick sensitizing films while maintaining high electron collection efficiency, which resulted in greatly improved performance (from 1.1% to 2.1%) compared to previously reported ZnO nanowire-based DSCs (Xu et al., 2010). J. X. Wang et al. used porous hierarchical disklike ZnO nanostructure fabricated via a simple low-temperature hydrothermal method as the photoanodes, and achieved improved photovoltaic performance (2.49%) compared with that of the ZnO nanowire arrays owing to the enlarged surface area and natural electron collection routes of ZnO hierarchical nanostructures (Wang et al., 2010). A. B. F. Martinson et al. introduced high surface area ZnO nanotube photoanodes templated by anodic aluminum oxide in DSCs via atomic layer deposition technique, providing a direct path for charge collection over tens of micrometers thickness (Martinson et al., 2007). Compared to similar ZnO-based devices, ZnO nanotube cells show exceptional photovoltage and fill factors, in addition to power efficiencies up to 1.6%. G Z Cao's group took the hierarchically popcorn-ball structured ZnO film as the photoanode, and attained the efficiency improvement from 0.6% to 3.5%, and finally 5.4%, by inducing light scattering within the photoelectrode films via the aggregation of ZnO nanocrystallites, thus demonstrating the huge potential of ZnO nanostructures in realizing DSCs with high efficiency (Chou et al., 2007; Zhang et al., 2008).

In parallel with photoanode films consisted by merely one type of material (ZnO, TiO_2, nanoparticles, nanowires, or nanotubes), the hybrid photoanode represents another important research trend, which is usually built by two or more materials or morphologies. Due to the obvious advantages of the hybrid photoanode including the improved interfacial structure/charge separation, the direct electron transporting path, and the vast possibilities to tune the microstructure and photoelectrochemical properties of the photoanode, this research trend has been gaining more and more interests. For example, K. M. Lee et al. incorporated multi-wall carbon nanotube in low-temperature fabricated TiO_2 photoanode, with the efficiency improvement of 20% (0.1% CNT), which may be resulted from retarded charge recombination between injected electrons and electron acceptors in the redox electrolyte (Lee et al., 2008). P. Brown et al. studied the single-wall carbon nanotube (SWCNT)/TiO_2 hybrid system, and proved that SWCNT support networks can be incorporated into mesoscopic TiO_2 films to improve the charge transport in DSCs (Brown et al., 2008). While no net increase in power conversion efficiency was obtained, an increase in photon-to-current efficiency (IPCE) represented the beneficial role of the SWCNT as a conducting scaffold to facilitate charge separation and charge transport in nanostructured semiconductor films. S. Pang et al. incorporated ZnO nanorods with different sizes into TiO_2 nanoparticle-based photoanode, and improved the conversion efficiency by 15% (Pang et al., 2007). They attributed the improved efficiency to the faster charge carrier transport rate, the decreased recombination, and the higher V_{oc}. B. Tan et al. used the composites of anatase TiO_2 nanoparticles and single-crystalline anatase TiO_2 nanowires as electrodes to fabricate DSCs, which possessed the advantages of both building blocks, i.e., the high surface area of nanoparticle aggregates and the rapid electron transport rate and the light scattering effect of single-crystalline nanowires (Tan et al., 2006). An enhancement of power efficiency from 6.7% for pure nanoparticle cells to 8.6% for the composite cell with 20 wt% nanowires was achieved, showing that employing nanoparticle/nanowire composites represented a promising approach for further improving the efficiencies of DSCs. C. H. Ku et al. reported ZnO nanowire/nanoparticle composite photoanodes with different nanoparticle-occupying extents (Ku et al., 2008). Aligned ZnO nanowires were grown on the seeded FTO substrate using an aqueous chemical bath deposition (CBD) first, and then, growth of nanoparticles among ZnO NWs by another base-free CBD was preceded further for different periods. The corresponding DSCs showed an efficiency of 2.37%, indicating the good potential of the hybrid nanostructures in ordered photoanodes.

Apart from the direct blending of two different semiconductor components as mentioned above, the coating technique has also been applied widely to create the hybrid photoanodes. M. Law et al. developed photoanodes constructed by ZnO nanowires arrays coated with thin shells of amorphous Al_2O_3 or anatase TiO_2 by atomic layer deposition (Law et al., 2006). They found that, while alumina shells of all thicknesses acted as insulating barriers that improve cell open-circuit voltage only at the expense of a larger decrease in short-circuit current density, titania shells in thickness of 10-25 nm can cause a dramatic increase in V_{OC} and fill factor with little current falloff, resulting in a substantial improvement in overall conversion efficiency (2.25%). They attributed the improved performance to the radial surface field within each nanowire that decreases the rate of recombination. K. Park et al. described a ZnO-TiO_2 hybrid photoanode by coating ultrathin TiO_2 layer by atomic layer deposition on submicrometer-sized aggregates of ZnO nanocrystallites (Park et al., 2010). The introduction of the TiO_2 ultrathin layer increased both the open circuit voltage and the fill factor as a result of the suppressed surface charge recombination without impairing the photocurrent density, thus realizing more than 20% enhancement in the conversion efficiency from 5.2% to 6.3%. S. H. Kang et al. examined effects of ZnO coating on the anodic TiO_2 nanotube array film on the conversion efficiency (Kang et al., 2007). Compared with the solid-sate cells consisted of an anodic TiO_2 film as the working electrode under backside illumination, an almost 20% improvement from the ZnO coating was achieved (from 0.578% to 0.704%), which can be attributed to the suppressed electron flow to the back-direction and the enhanced open-circuit voltage.

Despite considerable effects in this area, however, the record efficiency of 11% for DSCs is not surpassed by these new type cells, due to the complexity of both the nanoporous photoanode and the cell structure of DSCs. Much comprehensive and in-depth work related to this topic is required.

In this chapter, we focused on the ordered photoanode film built up by two semiconductor materials, zinc oxide (ZnO) and titanium oxide (TiO_2). Three type of ZnO nanostructures were selected, including the nanowire array (grown by the hydrothermal method), the nanoporous disk array grown on FTO substrate, and the nanoporous disk powder (transformed from the solution-synthesized zinc-based compound $ZnCl_2.[Zn(OH)_2]_4.H_2O$). Different types of TiO_2 nanoparticles were used, including commercial nanoparticles P25 & P90 (Degauss Co., Germany), and home-made hydrothermal TiO_2 nanoparticles, which have been widely used in producing traditional high-efficiency DSCs. Two kinds of preparation technique of ZnO-TiO_2 hybrid film were used according to the status of ZnO nanostructures (array or powder). The microstructure, optical and electrical properties of the hybrid film were investigated, and the performance of corresponding DSCs was measured and

Ordered Semiconductor Photoanode Films for Dye-Sensitized Solar Cells Based on Zinc Oxide-Titanium
Oxide Hybrid Nanostructures

87

compared with results of traditional cell. In special, the emphasis was placed on the controlling method of the microstructure of ZnO-TiO$_2$ hybrid films, and on the electron transporting mechanism in the hybrid films.

2. ZnO nanowire array/TiO$_2$ NPs hybrid photoanodes

In this section, two types of ZnO nanowire (NW) array were selected, i.e., dense and sparse NW array, with an aim to examine the effects of the distribution density of NW on the microstructure and photoelectrochemical properties of the hybrid cells. For the dense NW array, the ultrasonic irradition was used to promote the penetration of TiO$_2$ nanoparticles in the interstice of ZnO NWs.

2.1 Hybrid photoanodes based on dense ZnO NW array

ZnO nanowire (NW) arrays were grown on ZnO-seeded fluorinated tin oxide (FTO, 20 Ω/□) substrates by chemical bath deposition method. ZnO seed layer was prepared by sol–gel technique. ZnO NW arrays were obtained by immersing the seeded substrates upside-down in an aqueous solution of 0.025 mol/L zinc nitrate hydrate and 0.025 mol/L hexamethylenetetramine (HMT) in a sealed beaker at 90 °C for 12 h. After the deposition of ZnO NW, TiO$_2$ nanoparticles (NPs) were coated on ZnO NW by dipping the substrate into a well-dispersed TiO$_2$ suspension containing 0.5 g TiO$_2$ NPs (P25), 20 μL acetyl acetone, 100 μL Triton X-100 in 10 mL distilled water and 10 mL ethanol with 20 μL acetic acid. To facilitate the attachment and the gap filling of TiO$_2$ NPs into the interstices of ZnO NWs, the ultrasonic irradiation generated from a high-density ultrasonic probe (Zhi-sun, JYD-250, Ti alloy-horn, 20–25 kHz) was applied to TiO$_2$ suspension. The working mode was adjusted to work for 2 seconds and idle for 2 seconds, with the repetition of 99 cycles. The electrodes were then withdrawn at a speed of 3 cm per minute, dried, and sintered at 450 °C for 30 min in air. Figure 2 gave the schematic for the fabrication of the hybrid ZnO NW array/TiO$_2$ photoanopde.

For DSCs fabrication, ZnO NW based electrodes were immersed in a 0.5 mmol/L ethanol solution of N719 for 1 h for dye loading. The sensitized electrode was sandwiched with platinum coated FTO counter electrode separated by a hot–melt spacer (100 μm in thickness, Dupont, Surlyn 1702). The internal space of the cell was filled with an electrolyte containing 0.5 mol/L LiI, 0.05 mol/L I$_2$, 0.5mo/L 4-tertbutylpyridine, and 0.6 mol/L 1-hexyl-3-methylimidazolium iodide in 3-methoxypropionitrile solvent. The active cell area was typically 0.25 cm^2.

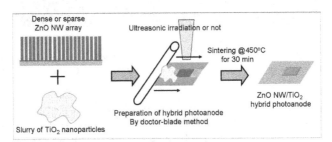

Fig. 2. Schematic of the preparation process of ZnO NW array/TiO$_2$ nanoparticles hybrid photoanode. NW: Nanowire.

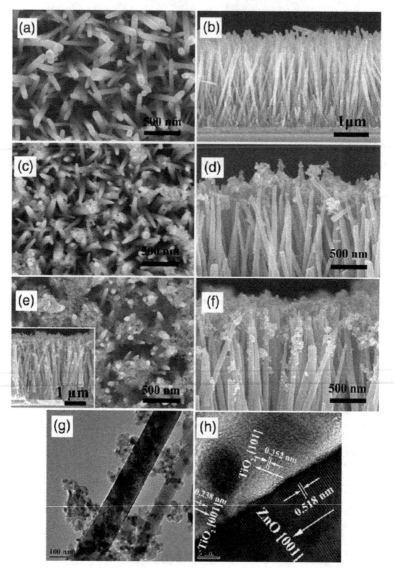

Fig. 3. FESEM images of ZnO NW arrays (a)–(b), hybrid ZnO NW/TiO₂ NP photoanodes prepared without (c)–(d), and with (e)–(f) the ultrasonic treatment. (g) Low and (f) high-resolution TEM images of the hybrid photoanodes prepared with ultrasonic treatment. (Reproduced from Ref. (Gan et al., 2007))

Figure 3 showed the top and side-view SEM images of ZnO NWs grown on FTO substrate and ZnO-TiO₂ hybrid photoanode film with/without ultrasonic treatment. Results indicate that, for ZnO nanowire array with a density of ~3.3×10⁹ cm⁻² and an average diameter of 80 nm and length of 3 μm, TiO₂ slurry with relatively high viscosity is difficult to penetreate into the inner pore of ZnO nanowires. As can be seen from Fig.3 c and d, only a small

Ordered Semiconductor Photoanode Films for Dye-Sensitized Solar Cells Based on Zinc Oxide-Titanium
Oxide Hybrid Nanostructures

89

amount of TiO_2 NPs were covered on the side surface of NWs and most of the NPs sit on the top of NWs without filling in the inner gaps. When the ultrasonic irradiation was applied, the coverage of NPs on the side surface of NWs was significantly improved (Fig. 3 e-h), and TiO_2 NPs were uniformly infiltrated into the interstices of NWs rather than stuck to the top of NWs. The cavitation in liquid–solid systems induced by the ultrasonic irradiation bears intensive physical effects, which can promote the transfer of TiO_2 NPs and drive them infiltrating into the gaps of NPs.

Figure 4 (left) showed the absorption spectra of the N719-sensitized ZnO NW, and hybrid ZnO NW/TiO_2 NP electrodes prepared with and without ultrasonic treatment, respectively. The absorption peak at around 515 nm, which corresponded to metal to ligand charge transfer (MLCT) in N719 dye (Nazeeruddin et al., 1993), significantly increased for the hybrid electrodes as compared to that of the pure ZnO NW electrode, proving that the dye-loading content is apparently increased upon the combination of ZnO NW with TiO_2 NPs. Besides, the hybrid electrode prepared with ultrasonic treatment showed an increase in the absorption in the wavelength range of 400–800 nm compared with that without ultrasonic treatment, indicating the higher surface area and the enriched light harvesting property by filling more TiO_2 NPs into the interstices between ZnO NWs with the assistance of ultrasonic irradiation.

Figure 4 (right) illustrated I–V characteristics of DSCs based on pure ZnO NWs and ZnO/TiO_2 hybrid photoanodes. Results show that the short-circuit current density (I_{sc}) and the conversion efficiency (η) of ZnO NWs based cell can be dramatically improved by incorporating TiO_2 NPs, which can be ascribed to the increase in the surface area and the dye loading quantity. However, the open-circuit voltage (V_{oc}) and the fill-factor (FF) of the hybrid DSCs decreased compared to those of pure ZnO NW DSC, which may be resulted from the increased interfaces and surface traps in the hybrid photoanode which may act as the recombination center under illumination. For the hybrid photoanode prepared with ultrasonic treatment, its I_{sc}, V_{oc}, FF, and η was 3.54 mA/cm^2, 0.60 V, 0.37, and 0.79%, respectively, indicating an approximately 35% improvement of the overall conversion efficiency compared with the photoanode without ultrasonic treatment. This improvement may originate from the enhanced light harvesting and the better attachment of TiO_2 NPs to ZnO NWs resulted from the efficient pore filling induced by the ultrasonic irradiation treatment.

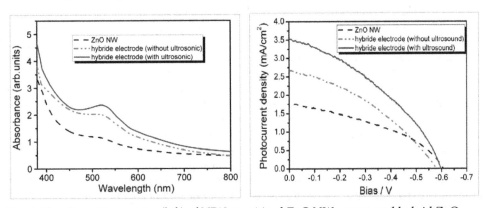

Fig. 4. The absorption spectra (left) of N719-sensitized ZnO NW arrays, and hybrid ZnO NW/TiO_2 NP photoanodes prepared without and with ultrasonic treatment, and I–V characteristics of corresponding DSCs (right). (Reproduced from Ref. (Gan et al., 2007))

In summary, these results indicate that, for the hybrid films combining dense ZnO NW array and TiO$_2$ NPs, the crucial aspect is to make TiO$_2$ NPs contained in the slurry penetrate into the deep interstice of ZnO NWs. The application of ultrasonic irradiation or other external fields may be helpful for the penetration of TiO$_2$ NPs, which usually result in the increase of the photoelectrochemical performance of the hybrid cells. However, it seems that the full filling of TiO$_2$ in the dense NW array is very difficult based on the current technique. So it is meaningful to develop the sparse nanowire array or other forms of TiO$_2$ NPs, to realize the good combination of ZnO NW array and TiO$_2$ NPs.

2.2 Hybrid photoanodes based on sparse ZnO NW array

In this section, ZnO NW array with sparse density was integrated with TiO$_2$ NPs, to form the hybrid photoanode. The growth of sparse ZnO NW array was realized by reducing the pH value of the precursor via the chemical bath deposition (CBD) method. The substrates and the experimental parameters were similar to those of dense one except the concentration of Zn^{2+} and HMT (both 0.02 mol/L), and the pH value (2.0-3.0).

TiO$_2$ slurry was prepared following the method in Ref (Ito et al., 2008), and the mass ratio of TiO$_2$, ethyl cellulose, and terpineol was 18 : 9 : 73. Due to the acid-dissolute nature of ZnO materials, the pH value of TiO$_2$ slurry should be controlled neutral or weak alkaline.

The preparation of the hybrid film based on sparse ZnO NW array was similar to that of dense array, as described in Section 2.1. The sensitization of the film was carried out in N719 dye solution dissolved in a mixture of acetonitrile and tertbutyl alcohol (volume ratio, 1:1) for 20-24 hours at room temperature. The fabrication of the cells was similar to the procedure described in Section 2.1, with the electrolyte composition of 0.6 M BMII, 0.03 M I$_2$, 0.10 M guanidinium thiocyanate and 0.5 M 4-tertbutylpyridine in a mixture of acetonitrile and valeronitrile (volume ratio, 85:15).

Figure 5 gave SEM images of sparse ZnO NW on the surface and cross section. It can be seen that the density of ZnO nanowire on FTO substrate is much sparser than the dense ZnO NW (Figure 3 a&b). But with the decrease of the density, the diameter of ZnO NW increases greatly, up to several micrometers.

The hybrid cell based on the sparse ZnO NW array exhibited the conversion efficiency of 2.16%, lower than the TiO$_2$ NPs-based cell (2.54%) as illustrated in Figure 6. The decreased efficiency of the hybrid cell is mainly resulted from the reduced photocurrent density compared with the TiO$_2$ cell, while the open voltage keeps unchanged and the fill factor improved from 0.06 to 0.078. The open-circuit voltage decay (OCVD) analysis (Figure 6) indicated that the hybrid film exhibits longer decay time when the illumination is turned off, indicating lower recombination rate between photo-induced electrons and holes. We believe that the obviously reduced photocurrent density may be related to the reduced surface area induced by the incorporation of large size ZnO nanowires, which may resulted in the reduced dye loading content. So the improvement in the efficiency of DSCs via the integration of sparse ZnO NW array and TiO$_2$ NPs is possible, as long as the size of ZnO nanowire can be reduced to tens of nanometers. However, limited by the current technology level of ZnO nanowire array, it is not an easy task to grow ZnO NW array both sparse and thin enough for the application in the hybrid photoanodes of DSCs.

In summary, we have successfully prepared the hybrid photoanode film using sparse ZnO

Ordered Semiconductor Photoanode Films for Dye-Sensitized Solar Cells Based on Zinc Oxide-Titanium
Oxide Hybrid Nanostructures

91

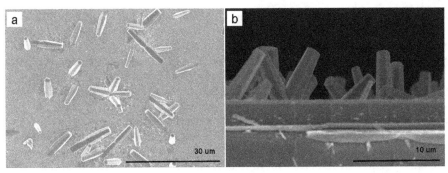

Fig. 5. FESEM images of sparse ZnO NW array on the surface (a) and the cross section (b).

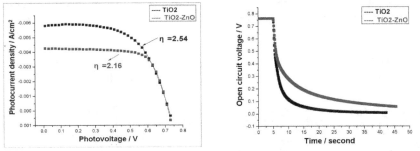

Fig. 6. I-V curves (left) and open-circuit voltage decay (OCVD) curves (right) of TiO_2 NPs-based cell and $ZnO-TiO_2$ hybrid cells based on sparse ZnO NW array under AM 1.5 illumination (100 mW/cm²). The active area is 0.27 cm² for all cells.

NW array and TiO_2 NPs. Although the total efficiency of the hybrid cell was lower than the TiO_2 NPs-based cell, the obvious improvement in the fill factor and the reduction in the recombination rate were observed. The reduced efficiency was mainly related to the decreased photocurrent density originated from the large-size ZnO NW. The further chance to improve the efficiency of ZnO NW based hybrid cell may reside in the realization of ZnO NWs with both sparse density and thin diameter.

3. ZnO nanoporous disk array/TiO_2 NPs hybrid photoanodes

In this section, an alternative ZnO nanostructure was used to prepare hybrid photoanode film, i.e., ZnO disk array possessing nanoporous feature. Compared with the traditional ZnO NW, the thickness of ZnO disk is lower and the surface area is higher. Thus higher effect in improving the conversion efficiency of DSCs can be expected.

ZnO nanoporous disk array was transformed from the disk of a layered zinc-based compound – simonkollite ($ZnCl_2.[Zn(OH)_2]_4.H_2O$, brief as ZHC) via calcinations. Conductive FTO glass coated by a thin TiO_2 layer (deposited by the hydrolysis of 40 mM $TiCl_4$ aqueous solution at 70oC) was used as the substrate. Typically, ZHC disk was prepared by CBD method. Aqueous solutions of 20 ml $ZnCl_2$ (0.2 mol/l), 20 ml hexamethylenetetramine (HMT) (0.2 mol/l), and 40 ml ethanol were mixed in a beaker and heated to 70oC in oven for 2 hours. After washing with H_2O and ethanol carefully, ZHC nanodisk array deposited on TiO_2/FTO substrate was sintered in air at 500oC for 4 hours, to convert ZHC to ZnO nanoporous disk.

TiO$_2$ NPs slurry was prepared by grinning TiO$_2$ commercial nanoparticles (P90, Degauss Co.) 0.5 g, H$_2$O 2.5g, PEG 20000 0.25 g in porcelain mortar. The ZnO-TiO$_2$ hybrid film was prepared by the doctor blade method, and the ZnO nanoporous disk array grown on TiO$_2$/FTO substrate was used. To achieve a specific thickness of the film, two layers of TiO$_2$ slurry were applied. The dried hybrid cell was sintered at 450°C in air for 30 minutes.

The sensitization of photoanode films and the fabrication of the cells were similar to those described in Section 2.1, except that the sensitizing time was prolonged to 24 hours.

Figure. 7(a) illustrated SEM images of ZHC nanodisk array deposited on TiO$_2$/FTO substrate. It can be seen that as-deposited ZHC exhibit rather regular hexagonal disk shape, with the size of ~ 10 um. The distribution of ZHC disks on substrate is sparse, satisfying the "low-content" requirement of ZnO in the hybrid photoanode film. After annealing at 500°C, ZHC disks were transformed into ZnO with typical nanoporous structure (as shown in Figure 7(b)), while the sheet structure (~100 nm in thickness) was maintained. Figure 7 (c) and (d) showed SEM images of the hybrid films based on this sparse nanoporous ZnO disk array. We can see that the morphology of the hybrid film on the surface and the cross section was rather smooth and uniform, with little difference from the traditional pure TiO$_2$ film (Gao, 2007). In addition, ZnO sheet like structures can not be found in either the surface or the cross section due to the low content of ZnO in the hybrid film.

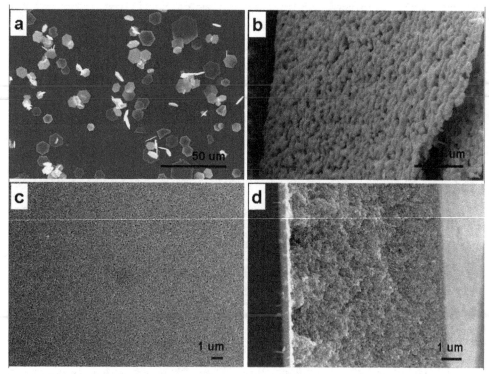

Fig. 7. FESEM images of (a) ZHC disk array and (b) ZnO nanoporous disk transformed from ZHC via calcinations at 500°C; FESEM images of ZnO-TiO$_2$ hybrid film based on sparse nanoporous ZnO disk array. (c) Surface and (d) cross section.

Figure 8 (left) gave the optical transmittance spectra of FTO substrate, pure TiO_2 film and the hybrid film. Results indicate that in the wavelength rage of 470-800 nm, the hybrid film possesses relatively lower transmittance than the pure TiO_2 film, while in the wavelength band of 300-470 nm, the transmittance of the hybrid film is higher. The reduced transmittance in the higher wavelength band of the hybrid film may be related to the scattering effects of the large ZnO disk in the film. In view of the maximum absorption of N719 dye molecules located at ~ 525 nm (Figure 4), the scattering of ZnO nanoporous disks to the incident light has positive influence on the performance of the hybrid cells. The reduced transmittance in the lower band of the pure TiO_2 film may be related to the increased agglomeration of TiO_2 NPs, which can induce the larger secondary particles and the higher scattering effects in the lower wavelength range. In contrast, the presence of large-size ZnO sheet may reduce the agglomeration phenomena to some extent, thus exhibiting higher transmittance.

Figure 8 (right) gave I-V curves of pure TiO_2 NPs cell and ZnO nanodisk array – TiO_2 NPs hybrid cell under AM 1.5 illumination (100 mW/cm²). It can be seen that the cell based on the hybrid film possesses much higher photocurrent density than TiO_2 NPs cell, increasing from 7.84 mA/cm² to 11.70 mA/cm². Also the improvements in the photovoltage and the fill factor of the hybrid cell are observed. As a result, the total conversion efficiency changes from 3.07% to 5.19%, increased by up to 60%.

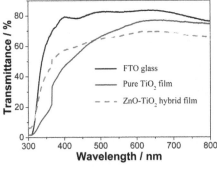 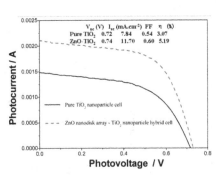

Fig. 8. The optical transmittance spectra (left) of pure TiO_2 NPs film and ZnO-TiO_2 hybrid film deposited on FTO substrate; I-V curves (right) of pure TiO_2 NPs cell and ZnO nanodisk array – TiO_2 NPs hybrid cell under AM 1.5 illumination (100 mW/cm²). The active area is 0.27 cm² for pure TiO_2 cell and 0.18 for ZnO-TiO_2 hybrid cell.

The reason for the efficiency improvement in the hybrid cell compared with NPs-based cell was analyzed by AC impedance under the illumination condition and open-circuit voltage decay (OCVD) analysis under the dark condition.

Figure 9 (left) showed Nyquist plots of the hybrid and pure photoanode, and the lower table gave the simulation results according to the physical model given in the inset. Two arcs can be clearly identified in the Nyquist plot for each sample. The left (high frequency) arc corresponds to the charge transfer process at the Pt counter electrode (R_{ct1}). The right large arc arises from the charge transport at the TiO_2/dye/electrolyte interface (R_{ct2}). The right small arc is related to the Warburg diffusion process of I^-/I_3^- in the electrolyte, which is not discussed in this work. The overall series resistance of the cell (R_s) is the resistance measured when electrons are transported through the device in the high-frequency range exceeding10[5]

Hz. By simulated calculation following the equivalent circuit, we can obtain the calculated value of R_s, R_{ct1}, and R_{ct2} for each sample. Results show that the hybrid film exhibits obviously lower R_s, R_{ct1}, and R_{ct2} than the pure TiO$_2$ film, indicating that the overall series resistance, the resistance at the Pt/electrolyte interface and at the TiO$_2$/dye/electrolyte interface in the hybrid cell is lower than the traditional TiO$_2$ NPs cell.

Figure 9 (right) showed OCVD curves of the hybrid and pure photoanode. While the pure TiO$_2$ cell exhibits rapid voltage decrease after the turning off of the illumination, the hybrid cell has much slower decay behavior, indicating that the photo-induced electron-hole recombination rate in the hybrid film is lower than the pure TiO$_2$ cell.

We believe the reduced overall resistance, the interfacial resistance and the electron-hole recombination rate is responsible for the obvious improvement in the total conversion efficiency in the hybrid cell.

In brief, we prepared ZnO-TiO$_2$ hybrid photoanode film based on sparse ZnO nanoporous disk array grown on TiO$_2$/FTO substrate. Though the obvious change in the microstructure of the film could not be observed, the hybrid film possessed increasing scattering effects in the wavelength range of 470-800 nm, which was beneficial to the light absorption of the dye molecules. Also the integration of ZnO nanoporous disk into TiO$_2$ NPs film resulted in the decrease of the overall series resistance and the resistance at the Pt/electrolyte interface and at the TiO$_2$/dye/electrolyte interface. As a result, the conversion efficiency was improved by 60%, indicating the great potential of the sparse ZnO nanoporous disk array in the field of hybrid DSC photoanodes.

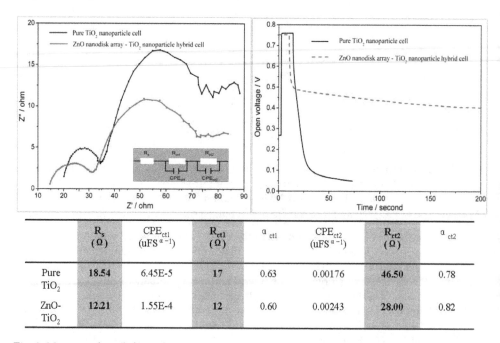

	R_s (Ω)	CPE_{ct1} (uFS$^{\alpha-1}$)	R_{ct1} (Ω)	α_{ct1}	CPE_{ct2} (uFS$^{\alpha-1}$)	R_{ct2} (Ω)	α_{ct2}
Pure TiO$_2$	18.54	6.45E-5	17	0.63	0.00176	46.50	0.78
ZnO-TiO$_2$	12.21	1.55E-4	12	0.60	0.00243	28.00	0.82

Fig. 9. Nyquist plots (left) and open-circuit voltage decay plots of pure TiO$_2$ NPs cell and ZnO nanodisk array – TiO$_2$ NPs hybrid cell. The attached table illustrates EIS parameters calculated from the given equivalent circuit.

Ordered Semiconductor Photoanode Films for Dye-Sensitized Solar Cells Based on Zinc Oxide-Titanium
Oxide Hybrid Nanostructures

95

4. ZnO nanoporous disk powder/TiO$_2$ NPs hybrid photoanodes

The disadvantage for the hybrid photoanode between ZnO array (both the nanoporous disk array and the nanowire array) and TiO$_2$ NPs lies in the difficulties in controlling the precise content of ZnO in the hybrid film, which is a crucial parameter for any composite material. Also, the distribution of ZnO array in the hybrid film may be not uniform, and difficult to control. Therefore, in this section, we attempted to blend ZnO nanoporous disk in the powder form into TiO$_2$ slurry, and prepared a uniform hybrid film via the doctor-blade technique. By this method, we can examine the effects of ZnO content in the hybrid film on the microstructure and properties of photoanode, and find an optimal composition for ZnO-TiO$_2$ hybrid photoanodes.

The powder of ZnO nanoporous disks was synthesized by the pyrolysis of chemical bath deposited ZHC nanodisks. Two types of TiO$_2$ NPs were selected, i.e., the commercial P25 TiO$_2$ and the hydrothermal TiO$_2$ NPs following the procedure described in Ref (Ito et al., 2008).

4.1 Hybrid photoanodes based on P25 TiO$_2$ NPs

Layered ZHC was prepared by the chemical bath deposition method. Typically, aqueous solutions of 20 ml ZnCl$_2$ (0.2 mol/l), 20 ml hexamethylenetetramine (HMT) (0.2 mol/l), and 40 ml ethanol were mixed together and heated to 70°C in oven for 2 hours, resulting in the white precipitation of ZHC. After centrifuging and drying, the powders were annealed at 500°C in a tube furnace in air for 18 hours to convert ZHC into ZnO.

Film electrodes for DSCs were deposited onto FTO substrate via TiO$_2$ or TiO$_2$/ZnO slurry by the doctor blade technique. For P25 TiO$_2$ NPs, the TiO$_2$ slurry was prepared using the mixed suspension of TiO$_2$ NPs (P25) (0.5 g) and PEG-1000 (0.25 g). The TiO$_2$/ZnO hybrid electrodes were made by using the mixed suspension of P25, PEG and ZnO disk powder, with the weight percentages of ZnO being 0 (S1), 0.5% (S3) and 1% (S2). The electrodes were sintered at 450°C for 120 min. The detail preparation process of the hybrid film was illustrated in Figure 10.

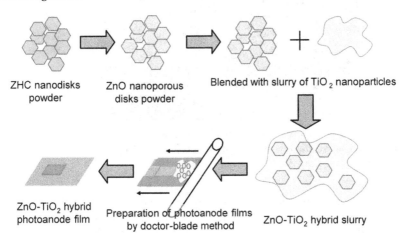

Fig. 10. Schematic of the preparation process of ZnO/TiO$_2$ hybrid photoanodes based on nanoporous ZnO disk powder and TiO$_2$ nanoparticles.

Fig. 11. SEM images of as-prepared ZHC powders and ZnO nanoporous disk after annealing at 500°C: (a) Low magnification morphology of ZHC powder; (b) Typical ZHC disks; (c) Low magnification morphology of ZnO nanoporous disks; (d) Enlarged view of nanoporous structure. (Reproduced from Ref (Gao et al., 2009))

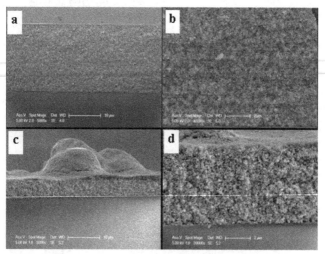

Fig. 12. SEM images of the cross section of TiO_2 film (a-b) and $ZnO-TiO_2$ hybrid film (c-d) prepared on glass substrate. (a) and (c): low magnification; (b) and (d): high magnification. (Reproduced from Ref (Gao et al., 2009))

The annealed electrodes were stained by N719 dye by soaking them in a 0.5 mmol/1 solution of N719 for 12 hours. FTO glass substrates were coated by Pt catalyst layer by decomposing $H_2PtCl_6 \cdot 6(H_2O)$ at 400°C, and were used as the counter electrode. The working electrode and the counter electrode were cohered together by surlyn 1702 hot melt foil. The electrolyte, consisted of 0.05 M I_2, 0.1 M LiI and 0.5 M tertbutylpyridine in acetonitrile, was filled into the cell from the hole in the counter electrode.

Figure 11 gave SEM images of as-prepared ZHC disks and nanoporous ZnO disks obtained by sintering. As-prepared ZHC possesses obvious disk-like feature, with the side length of 500-1000 nm and the thickness of 100-300 nm in average. After annealing, ZnO disks with the nanoporous feature were obtained, with the pore size ranging from 50-200 nm. In addition, the linking between neighboring ZnO particles in each disk can be clearly observed, which is expected to provide good electron transport in DSCs.

Figure 12 showed the cross-section morphologies of the pure TiO_2 film and 1%ZnO/TiO_2 hybrid film. While the pure TiO_2 film shows a uniform surface morphology and typical nanoporous structure with a thickness of ~20 μm, the hybrid film possesses a rough surface with large humps of ~10 μm (Fig.12 c) and lower thickness (~6 μm). The results indicate that even a small amount of ZnO powder blended in TiO_2 slurry can change the microstructure and thickness of the film electrodes significantly, which may be related to the change of the slurry viscosity during the preparation process. The presence of large ZnO particles (several μm in size) in the hybrid slurries may hamper the free flow of the TiO_2 slurry, thus resulting in the formation of large humps on the surface and the higher internal roughness

The optical transmittance spectra (Figure 13 (left)) of the TiO_2 and ZnO/TiO_2 hybrid films shows that both the pure and hybrid films exhibit strong scattering effects on the incident light in the visible and near infrared band. Compared with the pure TiO_2 film electrode, the hybrid film electrodes show much lower transmittance in the wavelength range of 500-1100 nm, indicating that a very small amount of ZnO disks can exert significant effects on the optical properties of the photoanode.

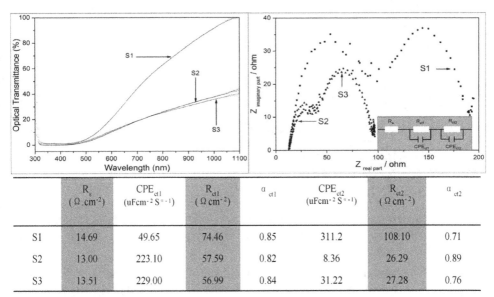

	R_s ($\Omega.cm^{-2}$)	CPE_{ct1} ($uFcm^{-2}S^{a-1}$)	R_{ct1} (Ωcm^{-2})	a_{ct1}	CPE_{ct2} ($uFcm^{-2}S^{a-1}$)	R_{ct2} (Ωcm^{-2})	a_{ct2}
S1	14.69	49.65	74.46	0.85	311.2	108.10	0.71
S2	13.00	223.10	57.59	0.82	8.36	26.29	0.89
S3	13.51	229.00	56.99	0.84	31.22	27.28	0.76

Fig. 13. Optical transmittance (left) of TiO_2 film (S1) and ZnO/TiO_2 hybrid film on FTO glass substrate, with ZnO percentage of 1% (S2) and 0.5% (S3), and electrochemical impedance spectra (right) of the DSCs based on TiO_2 electrode (S1) and ZnO-TiO_2 hybrid electrodes (S2 and S3). The attached table illustrates EIS parameters calculated from the given equivalent circuit. (Reproduced from Ref (Gao et al., 2009))

Figure 13 (right) showed the impedance spectra of DSCs using TiO_2 and ZnO/TiO_2 hybrid photoanodes. Results show that, both hybrid cells exhibit lower R_s, R_{ct1}, and R_{ct2} than those of the pure TiO_2 cell (S1), indicating that the incorporation of ZnO in the photoanode can decrease the overall series resistance of device significantly and facilitate the interfacial charge transport in both Pt/electrolyte and TiO_2/dye/electrolyte interface. The reason for this improvement may be the combination of the high electron transport nature of one-dimensional ZnO materials (Martinson et al., 2006), the large particle size and the network structure of ZnO disk.

Figure 14 revealed I-V curves of DSCs with TiO_2 and ZnO-TiO_2 hybrid films. The overall efficiencies of three cells are in the order of S3>S2>S1. While the cell using pure TiO_2 electrode (S1) exhibits the lowest efficiency of 1.1%, the cells with 0.5% and 1% ZnO-TiO_2 hybrid electrodes show higher efficiency of 2.7% and 2.3%, improved by 145% and 109%, respectively. Two hybrid cells exhibit similar V_{oc} but significantly higher I_{sc} and higher fill factor than pure TiO_2 cell, indicating that the improvement in the photocurrent density and fill factor is the main reason for the efficiency improvement. Also, the concentration of ZnO in the hybrid film should be no higher than 1% in this case, which is consistent with our previous observation in the hybrid film based on the sparse ZnO nanoporous disk array (Section 3) and results of other researchers on the DSCs with ZnO nanorod-TiO_2 hybrid electrode (Kang et al., 2007).

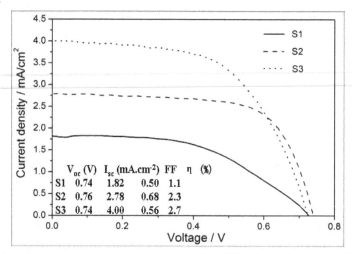

Fig. 14. The I–V characteristic curves of DSCs based on TiO_2 electrode (S1) and ZnO/TiO_2 hybrid electrode with ZnO percentage of 1% (S2) and 0.5% (S3). (Reproduced from Ref (Gao et al., 2009))

In summary, we have successfully demonstrated that ZnO nanoporous disk prepared from layered zinc based compound ZHC can be used to improve the efficiency of TiO_2 photoanodes effectively. The direct incorporation of ZnO nanodisk powder into TiO_2 slurry combined with the doctor-blade technique was proved an effective way to prepare the hybrid films. Results showed that even a small amount of ZnO incorporation in the TiO_2 film ($\leq 1\%$) can significantly influence the microstructure, optical and electrical properties. The rougher inner microstructure, the enhanced light-scattering effect on the visible and infrared light region, and the higher interfacial charge-transport rate were responsible for the improved efficiency in the hybrid photoanodes when compared with the pure TiO_2 film.

4.2 Hybrid photoanodes based on hydrothermal TiO_2 NPs

In this section, TiO_2 hydrothermal NPs were used as the source of TiO_2 for the preparation of the hybrid film. ZnO-TiO_2 hybrid slurry was prepared by adding specific amount of ZnO nanoporous disk powder into TiO_2 NPs slurry. TiO_2 NPs were prepared via the hydrothermal method and the slurry containing TiO_2 (18% by weight), ethyl cellulose (9%) and terpineol (73%) was prepared following Ref (Ito et al., 2008). ZnO nanoporous disk powder with the weight percentage of 0.5%-2% (compared with TiO_2) was blended into the slurry before the evaporation of ethanol via rotate-evaporator. Due to the acid-dissolute nature of ZnO materials, the pH value of TiO_2 slurry should be controlled at neutral or weak alkaline range.

ZnO-TiO_2 hybrid photoanodes were prepared by the doctor-blade technique using the hybrid slurry. Conductive FTO glass coated by a thin TiO_2 layer (deposited by the hydrolysis of 40 mM $TiCl_4$ aqueous solution at 70ºC) was used as the substrate. The sensitization of the photoanode film and the fabrication of the cells were similar to the procedure described in Section 2.2.

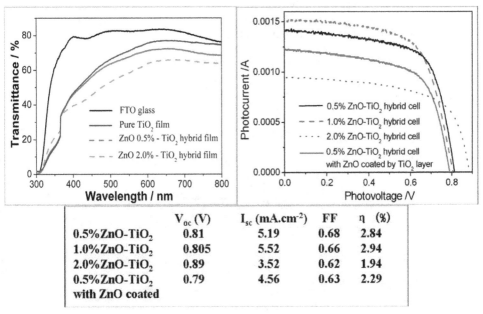

	V_{oc} (V)	I_{sc} (mA.cm^{-2})	FF	η (%)
0.5%ZnO-TiO_2	0.81	5.19	0.68	2.84
1.0%ZnO-TiO_2	0.805	5.52	0.66	2.94
2.0%ZnO-TiO_2	0.89	3.52	0.62	1.94
0.5%ZnO-TiO_2 with ZnO coated	0.79	4.56	0.63	2.29

Fig. 15. Optical transmittance (left) of ZnO-TiO_2 hybrid film deposited on FTO substrate, and I-V curves (right) of ZnO – TiO_2 hybrid cells based on different ZnO contents (0.5-2% by weight) under AM 1.5 illumination (100 mW/cm^2). The active area is 0.27 cm^2 for all cells.

SEM analysis (not shown here) indicates that the microstructure of the hybrid film was similar to that of pure TiO_2 film, different from the results in P25 slurry (Section 4.1). We think this difference may be resulted from the solvent of the slurry. In P25 based hybrid slurry, water is the dispersant, and the presence of large-size ZnO disk may exert significant influences on the viscosity of the slurry, thus modifying the microstructure of the hybrid film. In the hydrothermal TiO_2 hybrid slurry, organic terpineol was used as the dispersant, which possessed much higher viscosity than water. So the presence of a very small percentage of ZnO disks in the slurry had only minor influences on the viscosity and other physical properties of the hybrid slurry. Correspondingly, the microstructure of the hybrid film differed little from the pure one.

Figure 15 (left) showed optical transmittance spectra of ZnO-TiO_2 hybrid film deposited on FTO substrate. Different from microstructure results, the hybrid films exhibit obviously lower transmittance than the pure TiO_2 film in the visible-near infrared wavelength, which can be ascribed to the scattering effects of ZnO disks dispersed uniformly in the film. As mentioned above, this scattering behavior is beneficial to the absorption of N719 dye molecules. Also the hybrid film with 2% ZnO exhibits higher scattering effects than the one with 0.5% ZnO.

I-V curves given in Figure 15 (right) indicate that the photoanode with 1% ZnO possesses the highest efficiency compared with 0.5% ZnO and 2% ZnO hybrid films. While the cell parameters of the cells based on 0.5% and 1%ZnO are similar, the cell with 2%ZnO shows much less photocurrent density, which may account for the main fall in the efficiency. Also we were trying to coat a thin TiO_2 layer on ZnO nanoporous disk before its incorporation into TiO_2 slurry via the sol-gel method. Results show that this coating is not helpful in this case, maybe resulted from the reduced surface area in ZnO nanoporous disk due to TiO_2 coating. Further and in-depth investigation in this direction is required.

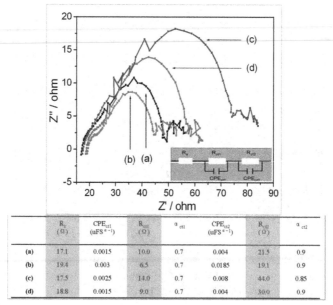

	R_s (Ω)	CPE_{ct1} (uFS^{a-1})	R_{ct1} (Ω)	a_{ct1}	CPE_{ct2} (uFS^{a-1})	R_{ct2} (Ω)	a_{ct2}
(a)	17.1	0.0015	10.0	0.7	0.004	21.5	0.9
(b)	19.4	0.003	6.5	0.7	0.0185	19.1	0.9
(c)	17.5	0.0025	14.0	0.7	0.008	44.0	0.85
(d)	18.8	0.0015	9.0	0.7	0.004	30.0	0.9

Fig. 16. Nyquist plots of ZnO nanodisk – TiO_2 NPs hybrid cells with ZnO content in hybrid slurry by weight of (a) 0.5%, (b) 1.0%, (c) 2.0%), and (d) 0.5% (ZnO disks were coated by a thin TiO_2 layer via sol-gel technique). The table illustrates EIS parameters calculated from the given equivalent circuit.

Ordered Semiconductor Photoanode Films for Dye-Sensitized Solar Cells Based on Zinc Oxide-Titanium
Oxide Hybrid Nanostructures

101

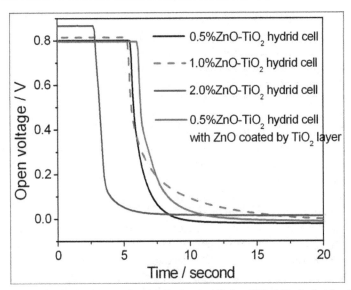

Fig. 17. Open-circuit voltage decay plots of ZnO nanodisk – TiO$_2$ NPs hybrid cells with
different ZnO content in hybrid slurry by weight (0.5%-2.0%).

Figure 16 illustrated Nyquist plots of ZnO-TiO$_2$ hybrid cells obtained from EIS analysis.
From the results of simulation, we can see that the hybrid cell with 1%ZnO content
possesses the lowest resistance at Pt/electrolyte interface and at TiO$_2$/dye/electrolyte
interface. The resistance at two interfaces increases when the content of ZnO in hybrid film
is either lower or higher than 1%. In special, for the hybrid film with 2% ZnO content, the
resistance at both interfaces is higher than that of film with 1% or 0.5% ZnO content. Also
for the film with 0.5% ZnO coated with a thin TiO$_2$ layer, the resistance at the
TiO$_2$/dye/electrolyte interface and the overall series resistance were increased, thus
resulted in the reduction in the total conversion efficiency. In general, we can derive a linear
relationship between the interfacial resistance obtained from EIS analysis and the overall
conversion efficiency of the hybrid cell.

Figure 17 showed OCVD plots of the hybrid cell with different ZnO content. Results show
that 1%ZnO-TiO$_2$ hybrid cell exhibits the slowest voltage decay rate among all four cells,
indicating the lower recombination rate, which is consistent with its high conversion
efficiency. Also the coating of a thin TiO$_2$ layer on ZnO nanoporous disk via sol-gel
technique can slightly improve the recombination properties of the hybrid cell, which is
beneficial to the improvement of the efficiency. However, other consequences induced by
the coating, possibly the blocking of the nanopore and the decrease of the surface area,
resulted in the reduction of the overall efficiency at last.

For ZnO-TiO$_2$ hybrid cell, it is important to discuss the effects of the treatment of the
photoanode film in TiCl$_4$ aqueous solution. ZnO is intrinsically an amphoteric oxide, which
can be dissolved in TiCl$_4$ aqueous solution with the strong acid nature. So it should be very
careful in treating ZnO-TiO$_2$ hybrid films in TiCl$_4$ solutions. Figure 18 illustrated I-V curves
of TiO$_2$ and ZnO-TiO$_2$ film with and without TiCl$_4$ treatment. Results show that the

incorporation of ZnO into TiO_2 can improve the efficiency from 3.5% to 3.98% before the $TiCl_4$ treatment. However, after the treatment, the efficiency of the hybrid film decreases to 3.2%, while the efficiency of the pure TiO_2 film increases to 6.38%. The dissolution of ZnO dispersed uniformly in the hybrid film may be underlying reason for the efficiency reduction. On the other hand, the $TiCl_4$ treatment is a very powerful method to improve the efficiency of the pure TiO_2 cell. So it is necessary to develop an alternative treatment method in a neutral aqueous solution, which has minor influence on the structure of ZnO. Other methods may be also helpful including the protective coating technique on ZnO nanoporous disks by thin TiO_2 layer which can resist the corrosion of acid. However, when the protective coating technique is adopted, one has to be careful in avoiding the blocking of the nanopore present in ZnO nanostructures.

In summary, this section discussed the preparation of ZnO-TiO_2 hybrid photoanode films based on ZnO nanoporous disk powder and hydrothermal TiO_2 NPs. An optimal ZnO content of ~1% in the hybrid film was observed, with the total conversion efficiency of ~2.94%. While the efficiency improvement in the hybrid cell was realized when compared with the pure TiO_2 film when no $TiCl_4$ treatment was performed, the decrease in the efficiency in the hybrid film after the $TiCl_4$ treatment indicated the sensitivity of the hybrid system. The development of other treatment methods suitable for ZnO-TiO_2 hybrid system or other protective means may afford the further direction in this field.

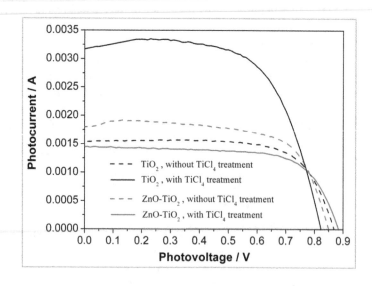

Fig. 18. I-V curves of TiO_2 NPs-based cell and ZnO–TiO_2 hybrid cells under AM 1.5 illumination (100 mW/cm²), illustrating the effects of $TiCl_4$ treatment.

Ordered Semiconductor Photoanode Films for Dye-Sensitized Solar Cells Based on Zinc Oxide-Titanium
Oxide Hybrid Nanostructures

103

5. Conclusion

In summary, the chapter started with a general review on the ordered TiO_2 and ZnO photoanode and the hybrid photoanode, outlining a brief picture on the status of the ordered photoanode in the field of DSCs. Then focusing on the ZnO-TiO_2 hybrid photoanodes, four type of ZnO nanostructure (including dense and sparse nanowire array, sparse nanoporous disk array, and nanoporous disk powder) and three type of TiO_2 nanoparticles (including P25, P90, and home-made hydrothermal nanoparticles) were used to prepare various hybrid films. Results show that, in general, the integration of ZnO with TiO_2 is a powerful means to improve the efficiency of the photoanode of DSCs, with the improvement up to 150%. However, one has to take great care in realizing the ideal hybrid structures. The content of ZnO in the hybrid film has to be maintained at a low level (e.g. ~1% by weight) in order to obtain a positive effect, which means that sparse and thin nanowire array or nanodisk array instead of the dense array is preferred. Also great care has to be taken during the $TiCl_4$ treatment of the ZnO-TiO_2 hybrid photoanode or the protective coating of ZnO nanostructures, preventing the destruction to the microstructure of ZnO nanostructures. Though at the current stage the overall conversion efficiency of the hybrid cell has not overpassed the highest level of the pure TiO_2 cell, it represents a very powerful and important technical route to optimize the microstructure and the performance of the ordered photoanode. We believe, by the continuous efforts of the forthcoming researchers, the hybrid semiconductor photoanodes with ordered structures will eventually shed the light to the DSCs with the breakthrough efficiency.

6. Acknowledgment

This work is supported by the 973-project (Grant no. 2009CB623304) of Ministry of Science and Technology of China and the Basic Research Program (Grant no. 51072214, 51002174) of National Natural Science Foundation of China.

7. References

Brown, P.; Takechi, K. & Kamat, P. V. (2008). Single-Walled Carbon Nanotube Scaffolds for Dye-Sensitized Solar Cells. *J. Phys. Chem. C*, Vol. 112, No. 12, pp. 4776-4782.

Chou, T. P.; Zhang, Q. F.; Fryxell, G. E. & Cao, G. Z. (2007). Hierarchically Structured ZnO Film for Dye-Sensitized Solar Cells with Enhanced Energy Conversion Efficiency. *Adv. Mater.*, Vol. 19, No. 18, pp. 2588-2592.

Gan, X. Y.; Li, X. M.; Gao, X. D.; Zhuge, F. W. & Yu, W. D. (2007). ZnO Nanowire/TiO_2 Nanoparticle Photoanodes Prepared by the Ultrasonic Irradiation Assisted Dip-coating Method. *Thin Solid Films*. Vol. 518, No. 17, pp. 4809–4812.

Gao, X. D.; Li, X. M.; Yu, W. D.; Qiu, J. J. & Gan, X. Y. (2007). Preparation of Nanoporous TiO_2 Thick Film and Its Photoelectrochemical Properties Sensitized by Merbromin. *J. Inorg. Mater.*, Vol. 22, No.6, pp. 1079-1085.

Gao, X. D.; Gao, W.; Yan, X. D.; Zhuge, F. W.; Bian, J. M. & Li, X. M. (2009). ZnO Nanoporous Disk-TiO$_2$ Nanoparticle Hybrid Film Electrode For Dye-Sensitized Solar Cells. *Funct. Mater. Lett.*, Vol. 2, No.1, pp. 27-31.

Grätzel M. (2001). Photoelectrochemical cells. *Nature*, Vol. 414, No. 6861, pp. 338-344.

Grätzel M. (2006). Photovoltaic Performance and Long-Term Stability of Dye-Sensitized Mesoscopic Solar Cells. *C. R. Chimie*, Vol. 9, No. 5/6, pp. 578-583.

Guldin, S.; Huttner, S.; Kolle, M.; Welland, M. E.; Muller-Buschbaum, P.; Friend, R. H.; Steiner, U. & Te´treault N. (2010). Dye-Sensitized Solar Cell Based on a Three-Dimensional Photonic Crystal. *Nano Lett.*, Vol. 10, No. 7, pp. 2303-2309.

Hosono, E.; Fujihara, S.; Honma, I. & Zhou, H. (2005). The Fabrication of an Upright-Standing Zinc Oxide Nanosheet for Use in Dye-Sensitized Solar Cells. *Adv. Mater.*, Vol. 17, No. 17, pp. 2091-2094.

Ito, S; Murakami, T. N.; Comte, P.; Liska, P.; Grätzel, C.; Nazeeruddin, M. K. & Grätzel M. (2008). Fabrication of Thin Film Dye Sensitized Solar Cells with Solar to Electric Power Conversion Efficiency Over 10%. *Thin Solid Films*, Vol. 516, No. 14, pp. 4613-4619.

Kang, S. H.; Kim, J. Y.; Kim, Y.; Kim, H. S. & Sung, Y. E. (2007). Surface Modification of Stretched TiO$_2$ Nanotubes for Solid-State Dye-Sensitized Solar Cells. *J. Phys. Chem. C*, Vol. 111, No. 26, pp. 9614-9623

Kang, S. H.; Choi, S. H.; Kang, M. S.; Kim, J. Y.; Kim, H. S.; Hyeon, T. & Sung, Y. E. (2008). Nanorod-Based Dye-Sensitized Solar Cells with Improved Charge Collection Efficiency. *Adv. Mater.*, Vol. 20, No.1, pp. 54 -58.

Ku, C. H.; Yang, H. H.; Chen, G. R. & Wu, J. J. (2008). Wet-Chemical Route to ZnO Nanowire-Layered Basic Zinc Acetate/ZnO Nanoparticle Composite Film. *Cryst. Growth Des.*, Vol. 8, No. 1, pp. 283-290.

Law, M.; Greene, L. E.; Johnson, J. C.; Saykally, R. & Yang P. (2005). Nanowire dye-sensitized solar cells. *Nature Materials*, Vol. 4, No. 6, pp. 455-459.

Law, M.; Greene, L. E.; Radenovic, A.; Kuykendall, T.; Liphardt, J. & Yang, P. (2006). ZnO-Al$_2$O$_3$ and ZnO-TiO$_2$ Core-Shell Nanowire Dye-Sensitized Solar Cells. *J. Phys. Chem. B*, Vol. 110, No. 45, pp. 22652-22663.

Lee, K. M.; Hu, C. W.; Chen, H. W. & Ho, K. C. (2008). Incorporating Carbon Nanotube in a Low-Temperature Fabrication Process for Dye-Sensitized TiO$_2$ Solar Cells. *Solar Energy Materials & Solar Cells*, Vol. 92, No. 12, pp. 1628-1633.

Li, B.; Wang, L. D.; Kang, B.; Wang, P. & Qiu, Y. (2006). Review of Recent Progress in Solid-State Dye-Sensitized Solar Cells. *Sol. Energy Mater. Sol. Cells*, Vol. 90, No. 5, pp.549-573.

Martinson, A. B. F.; McGarrah, J. E.; Parpia, M. O. K. & Hupp, J. T. (2006). Dynamics of Charge Transport and Recombination in ZnO Nanorod Array Dye-sensitized Solar Cells. *Phys. Chem. Chem. Phys.*, Vol. 8, No. 40, pp. 4655-4659.

Martinson, A. B. F.; Jeffrey, J. W.; Elam, W.; Hupp, J. T. & Pellin, M. J. (2007). ZnO Nanotube Based Dye-Sensitized Solar Cells. *Nano Lett.*, Vol. 7, No. 8, pp. 2183-2187.

Ordered Semiconductor Photoanode Films for Dye-Sensitized Solar Cells Based on Zinc Oxide-Titanium
Oxide Hybrid Nanostructures

105

Mor, G. K.; Varghese, O. K.; Paulose, M.; Shankar, K. & Grimes C. A. (2006a). A Review on
Highly Ordered, Vertically Oriented TiO_2 Nanotube Arrays: Fabrication, Material
Properties, and Solar Energy Applications. *Sol. Energy Mater. Sol. Cells*, Vol. *90*, No.
14, pp. 2011-2075.

Mor, G. K.; Shankar, K.; Paulose, M.; Varghese, O. K. & Grimes, C. A. (2006b). Use of
Highly-Ordered TiO_2 Nanotube Arrays in Dye-Sensitized Solar Cells. *Nano Lett.*,
Vol. 6, No. 2, pp. 215-218.

Nazeeruddin, M. K.; Kay, A.; Rodicio, I.; Humpbry-Baker, R.; Müller, E.; Liska, P.;
Vlachopoulos, N. & Grätzel, M. (1993). Conversion of Light to Electricity by Cis-
X2bis(2,2'-bipyridyl-4,4'-dicarboxylate) Ruthenium(II) Charge-Transfer Sensitizers
(X = Cl-, Br-, I-, CN-, and SCN-) on Nanocrystalline Titanium Dioxide Electrodes. *J.
Am. Chem. Soc.*, Vol. 115, No.14 , pp. 6382-6390.

O'Regan B. & Grätzel M. (1991). A Low-Cost, High-Efficiency Solar Cell Based on Dye-
Sensitized Colloidal TiO_2 Films. *Nature*, Vol. *353*, No. 6346, pp. 737-740.

Pang, S.; Xie, T.; Zhang, Y.; Wei, X.; Yang, M.; Wang, D. & Du, Z. (2007). Research on the
Effect of Different Sizes of ZnO Nanorods on the Efficiency of TiO_2-Based Dye-
Sensitized Solar Cells. *J. Phys. Chem. C*, Vol. 111, No. 49, pp. 18417-18422.

Park, K.; Zhang, Q.; Garcia, B. B.; Zhou, X. Y.; Jeong, Y. H. & Cao, G. Z. (2010). Effect of an
Ultrathin TiO_2 Layer Coated on Submicrometer-Sized ZnO Nanocrystallite
Aggregates by Atomic Layer Deposition on the Performance of Dye-Sensitized
Solar Cells. *Adv. Mater.*, Vol. 22, No. 21, pp. 2329–2332.

Rensmo, H.; Keis, K.; Lindstrom, H.; Sodergren, S.; Solbrand, A.; Hagfeldt, A. & Lindquist,
S. E. (1997). High Light-to Energy Conversion Efficiencies for Solar Cells Based on
Nanostructured ZnO Electrodes. *J. Phys. Chem. B.*, Vol. 101, No. 14, pp. 2598-
2601.

Rustomji, C. S.; Frandsen, C. J.; Jin, S. & Tauber, M. J. (2010). Dye-Sensitized Solar Cell
Constructed with Titanium Mesh and 3-D Array of TiO_2 Nanotubes. *J. Phys. Chem.
B*, Vol. 114, No. 45, pp. 14537–14543.

Tan, B. & Wu, Y. Y. (2006). Dye-Sensitized Solar Cells Based on Anatase TiO_2
Nanoparticle/Nanowire Composites. *J. Phys. Chem. B*, Vol. 110, No. 32, 15932-
15938.

Wang, J. X.; Wu, C. M. L.; Cheung, W. S.; Luo, L. B.; He, Z. B.; Yuan, G. D.; Zhang, W. J.; Lee,
C. S. & Lee, S. T. (2010). Synthesis of Hierarchical Porous ZnO Disklike
Nanostructures for Improved Photovoltaic Properties of Dye-Sensitized Solar Cells.
J. Phys. Chem. C, Vol. 114, No. 31, pp. 13157–13161.

Xu, C. K.; Shin, P.; Cao, L. L. & Gao, D. (2010). Preferential Growth of Long ZnO Nanowire
Array and Its Application in Dye-Sensitized Solar Cells. *J. Phys. Chem. C*, Vol. 114,
No. 1, pp. 125–129.

Zhang, Q.; Chou, T.; Russo, B.; Jenekhe, S. & Cao, G. (2008). Aggregation of ZnO
Nanocrystallites for High Conversion Efficiency in Dye-Sensitized Solar Cells.
Angew. Chem. Int. Edt., Vol. 47, No. 13, pp. 2402-2406.

Zhang, W.; Zhu, R.; Ke, L.; Liu, X. Z.; Liu, B. & Ramakrishna S. (2010). Anatase Mesoporous
TiO_2 Nanofibers with High Surface Area for Solid-State Dye-Sensitized Solar Cells.
Small, Vol. 6, No. 19, pp. 2176–2182.

Zhang Q. F. & Cao G. Z. (2011). Nanostructured Photoelectrodes for Dye-Sensitized Solar
        Cells. *Nano Today*, Vol. 6, No. 1, pp. 91-109.
Zhu, K.; Neale, N. R.; Miedaner, A. & Frank, A. J. (2007). Enhanced Charge-Collection
        Efficiencies and Light Scattering in Dye-Sensitized Solar Cells Using Oriented TiO_2
        Nanotubes Arrays. *Nano Lett.*, Vol. 7, No. 1, pp. 29-74.

5

Porphyrin Based Dye Sensitized Solar Cells

Matthew J. Griffith and Attila J. Mozer

ARC Centre of Excellence for Electromaterials Science and Intelligent Polymer Research Institute, University of Wollongong, Squires Way, Fairy Meadow, NSW, Australia

1. Introduction

Dye-sensitized solar cells (DSSCs) have emerged as an innovative solar energy conversion technology which provides a pathway for the development of cheap, renewable and environmentally acceptable energy production (Gledhill, Scott et al., 2005; O'Regan & Grätzel, 1991; Shaheen, Ginley et al., 2005). A typical DSSC consists of a sensitizing dye chemically anchored to a nanocrystalline wide band gap semiconductor, such as TiO_2, ZnO or SnO_2. The oxide structure is mesoporous in order to produce a high surface area for dye coverage, allowing the adsorbed monolayer to capture the majority of the incident solar flux within the dye band gap. The porous photoanode is immersed in an electrolyte which contains a redox mediator to transport positive charge to the counter electrode and maintain net electrical neutrality (Figure 1). Efficient charge separation is achieved through photoinduced electron injection from the excited state of the sensitizing dye into the conduction band of the metal oxide semiconductor. The resulting dye cations are subsequently reduced by the redox electrolyte, which also conducts the holes to the platinum-coated cathode. The solar to electric power conversion efficiencies of DSSCs depend on a delicate balance of the kinetics for injection, dye regeneration and recombination reactions (Haque, Palomares et al., 2005), with the best devices, currently based on ruthenium polypyridyl sensitizers and an iodide/triiodide redox mediator, exhibiting certified power conversion efficiencies of over 11% (Chiba, Islam et al., 2006).

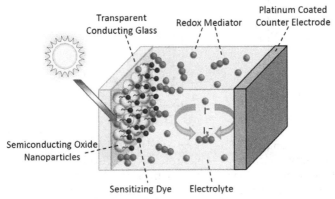

Fig. 1. Schematic illustration of a typical dye-sensitized solar cell (DSSC).

Porphyrin dyes have attracted significant interest as alternative sensitizers in DSSCs due to advantages with moderate material costs, ease of synthesis, large extinction coefficients and high stabilities. However, these dyes present a unique challenge since they have been found to possess very different operational photophysics to the majority of other sensitizing agents. Accordingly, porphyrin sensitizers create the opportunity to study some of the most fundamental limiting factors of DSSCs. In this chapter we will provide a detailed overview of the distinctive properties of porphyrin molecules and their behaviour as sensitizers in DSSCs. We focus on the major limitations affecting the performance of porphyrin DSSCs, including light harvesting, electron injection and charge recombination affects. We will also examine several strategies that have been employed to circumvent these limitations.

1.1 Operational principles of DSSCs

Unlike traditional silicon-based photovoltaic devices, charge separation and recombination in DSSCs are exclusively interfacial reactions. Furthermore, the initial photoexcited species in organic molecules are very different from those of silicon. Since organic molecules have lower dielectric constants and weaker Van der Waals interactions between molecules than their silicon counterparts, photoexitation of organic dyes produces a tightly bound neutral Frenkel exciton. This is in contrast to the loosely bound Werner excitons produced when silicon is photoexcited, which can essentially be considered free charges. Accordingly, DSSCs require an additional charge separation step to generate free charges. The viability of DSSCs for efficient photovoltaic energy conversion therefore relies almost entirely on achieving a delicate kinetic balance between the desired electron injection and dye cation regeneration reactions and the undesirable recombination reactions with either the dye cations or the acceptor species in electrolyte. The free energy driving forces for these various reactions are therefore crucial in determining the operational efficiency of DSSCs. These driving forces are often indicated by potential energy diagrams such as Figure 2, although such descriptions neglect entropy affects and thus do not strictly represent free energy.

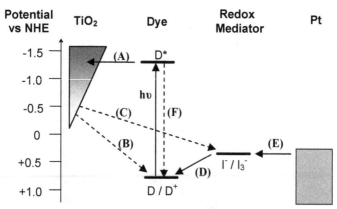

Fig. 2. Schematic representation of the energy levels of a DSSC indicating competing photophysical pathways, including (A) electron injection, (B) electron recombination with dye cations and (C) with the acceptor species in the electrolyte, (D) regeneration of dye cations by I-, and (E) recycling of I3- at the counter electrode. Figure taken from (Wagner, Griffith et al., 2011) and reproduced by permission of The American Chemical Society.

Current generation in the DSSC is dependent on three independent processes; the absorption of light by the photosensitizer, the injection of electrons from the excited photosensitizer, and the charge transport through the semiconductor film. The incident photon-to-current conversion efficiency (IPCE), also referred to as the external quantum efficiency (EQE), which corresponds to the electron flux measured as photocurrent compared to the photon flux that strikes the cell, is simply a combination of the quantum yields for these three processes as expressed in Equation 1.

$$IPCE(\lambda) = LHE(\lambda)\varphi_{inj}\,\eta_{coll} \tag{1}$$

Here LHE(λ) is the light harvesting efficiency for photons of wavelength λ, φ_{inj} is the quantum yield for electron injection and η_{coll} is the electron collection efficiency. The short circuit current density (J_{sc}) achieved by the device is simply the integrated overlap between the IPCE spectrum and the solar irradiance spectrum ($I_0(\lambda)$) over all wavelengths:

$$J_{sc} = \int q\,I_0(\lambda)\,IPCE(\lambda)\,d\lambda \tag{2}$$

The photovoltage generated by a DSSC is given by the difference in the Fermi energy, E_F, of electrons at the two contacts. Under electrochemical equilibrium in the dark, E_F must be equal for all components of the DSSC. Since the density of states in the semiconductor is not large enough to appreciably affect $E_{F,redox}$, the redox mediator Fermi level, the dark Fermi level is extremely close to $E_{F,redox}$. At open circuit and under illumination, the concentration of electrons in the TiO$_2$ film increases to a steady state value, n_{light}, determined by the balance of electron injection and recombination. The photovoltage, V_{photo} which corresponds to the increase in the electron Fermi level, is therefore determined by the ratio of the free electron concentration in the TiO$_2$ under illumination and in the dark:

$$V_{photo} = \frac{1}{q}\left(E_F - E_{F,Redox}\right) = \frac{K_B T}{q}\ln\frac{n_{light}}{n_{dark}} \tag{3}$$

The overall power conversion efficiency of a DSSC, η_{global}, is then determined from the intensity of the incident light (I_0), the short circuit current density (J_{sc}), the open-circuit photovoltage (V_{oc}), and the fill factor of the cell (FF) (which is simply the ratio of the maximum power obtained from a device to the theoretical maximum $J_{sc}V_{oc}$):

$$\eta_{global} = \frac{J_{sc}V_{oc}FF}{I_0} \tag{4}$$

The maximum value of η_{global} which can be obtained from a single junction solar cell is established as 32% (Shockley & Queisser, 1961), which accounts for photon absorption, thermalization, and thermodynamic losses encountered in converting the electrochemical energy of electrons into free energy to perform work. However, given the additional charge separation step required in a DSSC, a realistic efficiency limit is likely to fall well below this Shockley-Quiesser barrier due to restrictions on the allowable optical band gap (in order to maintain sufficient driving force for injection into TiO$_2$) and the significant loss of potential through the driving force required for regeneration of dye cations by the redox mediator.

If the semiconductor is chosen to be TiO$_2$, which to date is the only material to produce efficiencies over 10%, then the absorption onset limit at which the injection yield can still remain close to unity is currently observed to be around 900 nm for a ruthenium triscyanto terpyridyl complex ("black dye") (Nazeeruddin, Pechy et al., 2001). Beyond this limit a larger proportion of the solar flux can be harvested, but the reduction in the lowest unoccupied molecular orbital (LUMO) energy of the dye lowers the free energy driving force for electron injection, causing a subsequent reduction in the injection yield. Assuming a 10% loss of incident photons due to reflection from the top glass surface of the anode, an idealized IPCE spectrum with an absorption onset of 900 nm can be created by presuming a rise to maximum IPCE over ~50 nm and light harvesting, injection and collection efficiencies of unity for all wavelengths between 400 nm and 850 nm (a situation which can already be achieved with sensitizers where the absorption onset is 750 nm and which has almost been achieved with the black dye itself). The maximum possible photocurrent can then be calculated by integrating the overlap of this IPCE spectrum with the AM 1.5 solar irradiance spectrum (Figure 3a), and yields a value of 30 mAcm^{-2}.

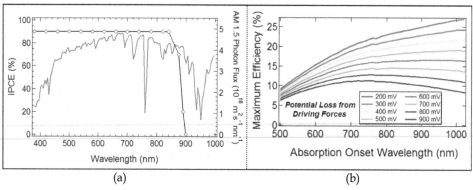

(a)   (b)

Fig. 3. (a) An idealized DSSC IPCE spectrum for a device with an absorption onset of 900 nm. The AM 1.5 solar irradiance spectrum (100 mWcm^{-2}) is also shown (red). (b) The maximum DSSC efficiency obtained from various absorption onsets with several possible potential energy losses from electron injection and dye regeneration driving forces.

To compute the maximum photovoltage, the free energy driving forces necessary to drive both electron injection and dye cation regeneration with yields of close to unity must be known. The sum of these driving forces is then removed from the optical band gap of the dye to calculate the maximum obtainable V_{oc}. There is a general lack of understanding regarding the minimum required driving forces, although previous studies indicate that the driving force required to achieve quantitative electron injection is ~200 mV (Hara, Sato et al., 2003; Katoh, Furube et al., 2002). The driving force required for dye regeneration will depend on the redox mediator, and is unusually high (~500 mV) for the commonly employed I$^-$/I$_3^-$ system as this is a two electron process (Boschloo & Hagfeldt, 2009). If we assume that the regeneration driving force can be reduced to approximately 350 mV with an alternative redox mediator which involves only a single electron transfer, then the loss of potential from injection and regeneration driving force requirements would be 550 mV. Including the photon absorption, thermalization and energy conversion losses inside these driving force potentials, a 550 mV loss of potential would yield a maximum V_{oc} of 0.83 V.

Calculating the fill factor of a DSSC remains an ambiguous task. Bisquert et al. reported that the Shockley diode equation can be employed to calculate the fill factor as a function of the photovoltage. However, this approach requires knowledge of the diode ideality factor, which is still an unclear value for DSSCs (Bisquert & Mora-Sero, 2010). Since this is a matter of debate in the field, for the current calculations we have assumed a value of 0.70, which is either achieved or surpassed in many of the current benchmark devices. From equation (4), it is then possible to compute the maximum theoretical efficiency for such a DSSC. Since there will inevitably be debate about the acceptable assumptions for the maximum absorption onset and minimum driving forces for injection and regeneration, we have also computed the theoretical efficiency for several alternative values using a fill factor of 0.70 (Figure 3b). These values range from 14.5% (absorption onset 800 nm, loss of potential 700 mV) to 21.3% (absorption onset 950 nm, loss of potential 400 mV). Using realistic assumptions for absorption onset (900 nm) and loss of potential (550 mV), we calculate the maximum practical efficiency for a DSSC to be 17.6%. Considering the highest reported certified efficiency for a DSSC is 11.1%, there are clearly many limitations which still affect these devices and must be removed in order to approach this practical efficiency limit.

1.2 Porphyrin sensitizers in DSSCs

Emulation of the extraordinary chlorophyll-based photosynthetic light harvesting apparatus has inspired researchers to investigate synthetically prepared porphyrin dyes as sensitizers for dye-sensitized light harvesting applications (Kay & Grätzel, 1993), and they remain one of the most frequently studied dyes (Campbell, Burrell et al., 2004; Campbell, Jolley et al., 2007; Imahori, 2010). They are attractive for such purposes as their synthesis is relatively straightforward and their optical and electronic properties can be tuned via chemical modification of the porphyrin core (Dos Santos, Morandeira et al., 2010), the number of porphyrin units (Mozer, Griffith et al., 2009), and the linker between the core and the inorganic oxide (Lo, Hsu et al., 2010). It was originally assumed that porphyrin dyes would function identically to other analogous chromophores inside a DSSC, however, extensive studies of porphyrin dyes have shown that they are distinctive, and possess very different photophysics to the majority of other sensitizers (Bessho, Zakeeruddin et al., 2010; Campbell, Jolley et al., 2007; Imahori, Umeyama et al., 2009). The structure of some of these key porphyrin sensitizers are shown in Figure 4. Consequently, porphyrin dyes offer a unique opportunity to understand some fundamental limitations of DSSCs and trial innovative strategies which manipulate their unusual photophysical properties.

Porphyrin molecules typically display several strong visible light absorption bands due to the π-π^* electron transitions of the macrocycle. Optical transitions from the two closely spaced highest occupied molecular orbitals (HOMO and HOMO + 1) to degenerate lowest unoccupied molecular orbitals (LUMOs) interact strongly to produce a high energy S_2 excited state with a large oscillator strength around 420 nm (Soret band), and a lower energy S_1 excited state with diminished oscillator strength between 500 nm and 650 nm (Q-band). Several Q-band absorptions are normally observed due to the vibronic transitions of the S_1 excited state, with the exact number depending on the symmetry of the porphyrin core (Gouterman, 1978). The position of these absorption bands can be tuned by structural modifications to the porphyrin molecule, and thus provide an extremely promising method of maximizing absorption of incident photon flux.

Porphyrin photoluminescence spectra display very small Stokes shifts and appear as two peaks due to transitions from different vibronic levels of the S_1 state back to the S_0 ground state. Interestingly, metalloporphyrins, and in particular many zinc porphyrins, can display dual fluorescence, that is, emission from both the S_2 and S_1 excited states, due to the large energy gap between S_2 and S_1 excited states causing slow internal conversion. The relaxation dynamics of excited porphyrin complexes are well established. Singlet excited state lifetimes are normally ~1 ps from the S_2 state and ~2 ns from the S_1 state for metalloporphyrins and ~10ns from S_1 for free base (metal free) molecules. Electrochemically porphyrin molecules are quite stable, with metalloporphyrins generally displaying two reversible oxidations of the porphyrin ring to form a radical cation or dication, while free base molecules generally show at least a single reversible oxidation. The assignment of the free base species is complicated by the stability of the mono and dications being strongly influenced by the identity of the solvent (Geng & Murray, 1986). The typical absorption, photoluminescence and electrochemical properties of porphyrin dyes are illustrated in Figure 5.

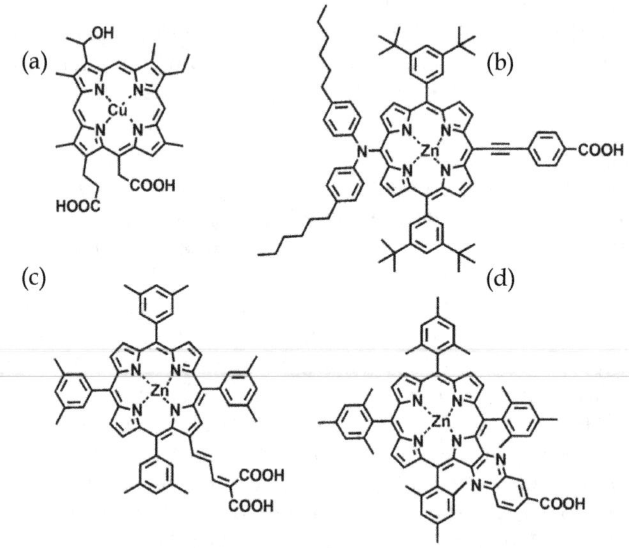

Fig. 4. The structure of (a) a pioneering synthetic porphyrin dye, and highly efficient porphyrin dyes (b) YD2, (c) GD2 and (d) Zn-4 published by various groups.

The development of new porphyrin dyes has been coupled with an increased understanding of the photophysics in operational DSSCs. For example, through various collaborations our group has previously shown that the luminescence lifetime, indicative of charge injection, depends on both the thermodynamic free energy driving force for injection and the conjugation through the linker moiety (Dos Santos, Morandeira et al., 2010). Furthermore, the electron lifetime in two of the most efficient porphyrin-sensitized solar cells was shown to be an order of magnitude lower than in identically prepared state of the art ruthenium bipyridyl (N719)–sensitized solar cells (Mozer, Wagner et al., 2008). This limitation can be somewhat circumvented by various post-treatments of the porphyrin-sensitized solar cells (Allegrucci, Lewcenko et al., 2009; Wagner, Griffith et al., 2011). However, despite this advanced understanding, the power conversion efficiencies of the best porphyrin-sensitized solar cells remains around 11%, well short of the theoretical maximum calculated earlier. The remainder of this chapter will focus on the fundamental limitations of these devices and explore some strategies which have been implemented to circumvent these limitations.

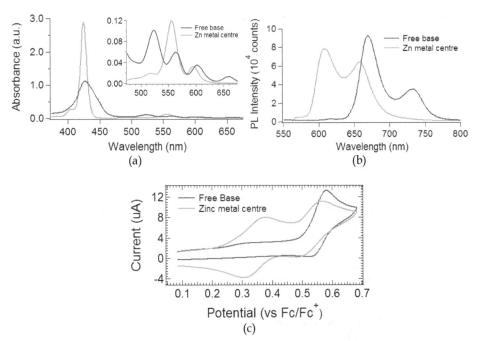

Fig. 5. The typical photophysical properties for 5×10^{-6} M solutions of zinc (green) and free base (brown) porphyrin molecules, showing (a) UV-vis absorption, (b) photoluminescence and (c) cyclic voltammograms (supporting electrolyte = 0.1 M TBAP).

2. Charge generation

The generation of free charges is the first critical step in achieving high efficiency in DSSCs. As indicated by equation (1), the charge generation processes consists of three steps; light absorption by the dye, injection of electrons from the photoexcited dye, and charge transport through the semiconductor film to prevent recombination. There are fundamental limitations with each of these three processes for porphyrin-sensitized solar cells. Some of these issues are intrinsic to the sensitizer and are difficult to remove. However, many of these basic problems can be circumvented using innovative approaches to device design. In this section we discuss some of the major limitations with charge generation in porphyrin-sensitized solar cells and summarize some of the important strategies that can be employed to reduce or indeed remove these limitations.

2.1 Light harvesting efficiency

Since the cross section for photon absorption of most photosensitizers is much smaller than the geometric area occupied on the semiconductor surface, light absorption by a pigment monolayer is small (Grätzel, 2005). To circumvent this, nanostructured semiconductor electrodes with a surface roughness factor (internal surface area normalized to the geometric area) on the order of a 1000 have been used, however these still require semiconductor films of at least 6 μm to produce a light harvesting efficiency close to unity. The requirement for this semiconductor thickness places significant constraints on the design of the DSSC. One limitation is that dyes which are prone to recombination, such as porphyrin sensitizers, suffer increasing limitations with thicker semiconductor films due to the larger surface area for interfacial electron transfer reactions such as recombination. Furthermore, the viscosity of the electrolyte must be kept low to enable complete filling of the nanopores throughout the entire film thickness. Since such solvents are typically organic, their volatility leads to problems with hermetic sealing and long term stability of devices.

Incorporating multichromophore light harvesting arrays with increased absorption cross sections could potentially solve both of these issues, enabling efficient DSSCs with thinner films using ionic liquid (Kuang, Ito et al., 2006) or solid state electrolytes (Bach, Lupo et al., 1998). Multichromophore dyes could also allow novel electrode structures with smaller internal surface areas, such as nanotubes (Mor, Shankar et al., 2006), nanowires (Martinson, Elam et al., 2007), or large porosity mesoscopic structures to be employed. Our group investigated such an approach by joining two identical porphyrin chromophores together to create a dimer with twice the extinction coefficient but an unchanged molecular footprint on the semiconductor surface (Figure 6a). However, to ensure the enhanced light harvesting results in increased charge generation, efficient electron injection from both chromophores must be achieved. This condition is extremely difficult to demonstrate for dyes employing identical chromophores due to the matching spectroscopic features of each unit making them impossible to distinguish. In collaboration with Japanese co-workers we were able to produce the first spectroscopic evidence for electron injection from both chromophores of this porphyrin dimer by employing femtosecond time-resolved transient absorption (Mozer, Griffith et al., 2009). Since photons at the probe beam wavelength of 3440 nm are primarily absorbed by electrons in the TiO₂ in the absence of dye aggregates (Furube, Katoh et al., 2005), electron injection from the photoexcited dye is directly monitored by this technique. Furthermore, after normalization of the signal to correct for the absorbed pump beam intensity, a comparison of both the injection yield and kinetics can be performed. With reference to an N719 standard dye reference to determine the 100% injection yield value, the injection yield of the porphyrin dimer was determined to be ~70% (Figure 6b). The >50% quantum yield for the porphyrin dimer clearly demonstrates electron injection from *both* photoexcited porphyrin units within the dimer.

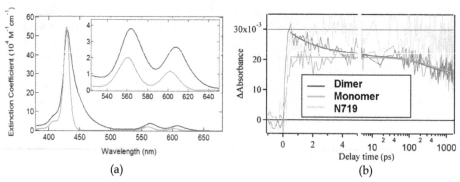

(a)   (b)

Fig. 6. Molar extinction coefficients of a porphyrin monomer (green) and dimer (blue) in solution. (b) Fs-TA signals (corrected for the absorbed pump intensity) of porphyrin-sensitized TiO₂ films in a redox containing electrolyte. An N719 signal obtained in air is also shown. Samples were excited by 150 fs pulses at 532 nm. Figure 6b taken from (Mozer, Griffith et al., 2009) and reproduced by permission of The American Chemical Society.

A second issue with light harvesting using porphyrin molecules is that the absorption bands of the sensitizers are quite discrete. Whilst the extinction coefficients in these absorption bands can be very high by employing for instance, a multichromophore dye, there are significant areas of the solar irradiance spectrum that are not covered by the absorption spectrum. Most notably, the vast majority of porphyrins do not absorb light beyond 700 nm. Additionally, the region between the Soret and Q-bands (450 nm-550 nm) also has a low absorption of incident photons. This lack of absorption significantly constrains the photocurrent which can be obtained from a porphyrin-based DSSC. Several strategies have been explored in an attempt to harvest an additional proportion of the incident solar flux using porphyrin dyes. One approach is to explore synthetic routes to extend the sensitizer absorption further towards the red end of the spectrum. Extension of the π conjugation and removal of symmetry in the porphyrin core can lead to splitting of the π and π^* levels and a decrease in the HOMO-LUMO gap, resulting in significant broadening and a red shift of the absorption bands. Imahori et al. were able to employ such an approach by using π-extended and fused porphyrin dyes to enhance the IPCE and overall device efficiency in a series of porphyrin molecules (Imahori, Umeyama et al., 2009).

A second approach used to extend the light harvesting efficiency of porphyrin-sensitized solar cells is to co-sensitize the semiconductor electrode with additional dyes which have complimentary absorption spectra to the porphyrin (Figure 7).

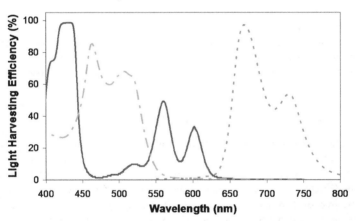

Fig. 7. An example of the extension in the porphyrin light harvesting efficiency (purple) that can be achieved by adding rhodium complex (green) and phthalocyanine (orange) dyes.

This approach has been utilized for many types of dyes, and was recently employed by Bessho et al. to enhance the photocurrent extracted from DSSCs prepared with a zinc porphyrin dye (Bessho, Zakeeruddin et al., 2010). The co-sensitization approach allows the absorption spectrum to be successfully extended; however, given the finite number of dye binding sites available on the semiconductor surface, there is still an intrinsic limitation on overall light harvesting potential. Our group explored methods to circumvent such limits by investigating a model system which combined free base and zinc porphyrin dyes. This study utilized not only the complimentary absorption spectra of the two dyes but also examined other possible synergistic interactions between the two dyes which are necessary in order to overcome the limit of a fixed amount of dye binding sites on the surface. We were able to demonstrate an enhancement in both light harvesting *and* the injection yield when zinc and free base porphyrin dyes were combined on the same TiO_2 surface (Griffith, Mozer et al., 2011). Other groups have also pursued similar studies, focusing on extending the co-sensitization concept using energy relay systems. This approach involves dissolving the co-sensitizer in the electrolyte so that it no longer competes with the major sensitizer for binding sites on the semiconductor. Absorbed photon energy is transferred from the dissolved co-sensitizer to the chemically bound major sensitizer where it is then injected into the semiconductor. This approach achieved photocurrent enhancements of ~30% compared to direct co-sensitization on the same semiconductor surface (Hardin, Hoke et al., 2009).

2.2 Electron injection into semiconducting oxides

Electron injection from the photoexcited dye into the acceptor states of the semiconductor conduction band is perhaps the key mechanistic step in achieving efficient charge generation in DSSCs. According to the classical theory of electron transfer developed by Marcus, the rate of electron transfer, k_{ET}, between discrete donor and acceptor levels under non-adiabatic conditions is given by (Marcus, 1964):

$$k_{ET} = \frac{2\pi}{\hbar} \frac{H^2}{\sqrt{4\pi\lambda k_B T}} \exp\left(\frac{-\left(\lambda + \Delta G_0{}^2\right)}{4\pi\lambda k_B T}\right) \qquad (5)$$

where H^2 is the electronic coupling between donor and acceptor states, ΔG_0 is the free energy driving force for electron transfer, λ is the total reorganization energy, T is the absolute temperature and h and k_B the Planck and Boltzmann constants respectively. The electronic coupling (H^2) decreases exponentially with increasing distance, d, between the donor and the acceptor as:

$$H^2 = H_0{}^2 \exp\left(-\beta d\right) \qquad (6)$$

where β is related to the properties of the medium between donor and acceptor, and $H_0{}^2$ is the coupling at distance d = 0. To achieve high efficiencies for injection in DSSCs, electron injection must be at least an order of magnitude faster than the competing deactivation of the dye excited state. Extensive studies of this charge separation process have typically shown sub-ps injection dynamics, suggesting electron injection competes efficiently with excited state decay, which occurs on the 1-10 ns timescale for porphyrin dyes. However, despite such fast kinetics, many porphyrin dyes still show very poor injection efficiencies. One possible reason for this poor injection is the heterogeneous nature of the process. Koops and Durrant demonstrated a distribution of injection half-life time constants from 0.1 – 3 ns for devices sensitized with various ruthenium polypyridyl dyes. They attributed this result to variations in the local density of acceptor states in the semiconductor for electron injection and therefore in the integrated electronic coupling, H^2, for this reaction (Koops & Durrant, 2008). Since such behaviour is dependent on the density of states in the semiconductor and not on the dye itself, it would seem acceptable to assume that such heterogeneous injection kinetics also apply to porphyrin dyes, and thus there may be some slow injecting dyes which cannot compete with excited state deactivation.

The structure of the dye is clearly one crucial factor which will determine the injection efficiency. Campbell et al. investigated a wide range of porphyrin dyes and discovered that the binding group which provides the electronic linkage between the chromophore and the semiconducting oxide plays an important role on the extracted photocurrent of devices. Given the similarity in the overall dye structures tested, this difference was attributed to variations in the injection efficiency achieved by varying the electronic coupling with different binding groups. Furthermore, the position of the binding group with respect to the porphyrin ring also affected the injection efficiency, with β-pyrollic linked groups showing better efficiency than meso linked groups. Our group extended such investigations in collaboration with co-workers in England. It was shown using luminescence quenching coupled with time correlated single photon counting detection to probe injection, that both the conjugation in the linker moiety and the metallation of the porphyrin can affect the injection yield in porphyrin systems (Figure 8). Peripheral substituents in the meso positions of the porphyrin core have also been shown to effect injection, with bulky groups (phenyl or tert-butyl) providing steric hindrance effects which reduces dye aggregation or electron donating groups affecting the HOMO–LUMO gap of the dye and thus the driving force for injection (Lee, Lu et al., 2009).

	FbNC[a]	FbC[a]	ZnNC[a]	ZnC[a]	ZnC[b]
τ_{Zr} / ns	7.5	4.2	1.5	1.5	0.8
τ_{Ti} / ns	4.8	2.5	0.8	<0.1	0.6
$(k_{inj})^{-1}$ / ns	14	6.2	1.9	<0.1	1.7
J_{SC} / mAcm^{-2}	0.11	0.6	1.2	4.3	4.3

[a] Kinetics for films covered with propylene carbonate. [b] Films covered with electrolyte.

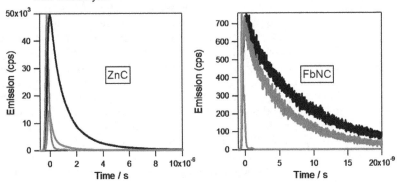

Fig. 8. (Top) Emission decay lifetimes, injection rate constants and device photocurrents for a series of porphyrin dyes with different metallation and linker conjugations. (Bottom) Transient emission decays of (a) a zinc porphyrin with a conjugated linker, and (b) a free base porphyrin with a saturated benzoic acid linker. Both dyes are adsorbed to TiO$_2$ (red), and ZrO$_2$, a high band gap semiconducting oxide which prevents electron injection (black). The instrument response function (IRF) is shown in grey. Figure taken from (Dos Santos, Morandeira et al., 2010) and reproduced by permission of The American Chemical Society.

Another concept which has been applied to improve injection in DSSCs is to synthesize dyes with an electron acceptor component close to the TiO$_2$ and an electron donor component furthest from the TiO_2 linker. This ensures the electron density in the excited state is concentrated in the vicinity of the TiO_2, promoting injection and localizing the resultant positive charge away from the interface, thereby reducing recombination. Given the ease with which porphyrin compounds can be synthetically modified, this class of dyes offers an ideal system to explore this donor–acceptor concept. Clifford et al tested the theory by modifying a zinc porphyrin with a triphenylamine electron donor, and showed that recombination of the injected electron with the dye was an order of magnitude slower than for a comparable dye that lacked the electron-donor groups (Clifford, Yahioglu et al., 2002). Hsieh et al extended such investigations when they tested a comprehensive range of electron donors and acceptors attached to the same porphyrin core. They demonstrated that several different electron donors attached to the optimal position of the porphyrin core were able to increase both the J_{sc} and the V_{oc} of the DSSCs, attributing this result to improved electron injection and reduced recombination due to the localization of electron density in the dye upon photoexcitation (Hsieh, Lu et al., 2010).

From equation (6) it is clear that the electronic coupling, and thus the rate of electron transfer for injection, is strongly dependent on the distance over which electron transfer occurs. If transfer between the porphyrin core and semiconductor occurs through the connecting binding group, extending the length of this group should reduce the speed with which injection occurs. Imahori et al tested this concept in a range of zinc porphyrin dyes, and found that contrary to expectation, the electron transfer process for longer linking groups were accelerated. They rationalized this result by postulating that some fraction of the porphyrin molecules are bound at an angle to the semiconductor surface as the linker becomes longer, with electron transfer in these dyes occurring through space, without facilitation through the linker. According to classical tunnelling theory, without the enhanced electronic coupling provided by the linker group, through-space injection could only occur if the sensitizer is within ~1 nm of the semiconductor surface. A distribution of electronic couplings from different injection routes would help explain the observed heterogeneity of the injection rates in DSSCs, however, dye orientation information remains quite limited. This lack of knowledge is problematic since the surface orientation of dyes will strongly affect the functioning of DSSCs, altering the effective barrier width for through-space charge tunnelling (Hengerer, Kavan et al., 2000) or the alignment of the dipole moment of the dye (Liu, Tang et al., 1996), which in turn can influence injection and recombination (Figure 9a). Several measurement techniques have been trialled, such as near edge X-ray absorption fine structure measurements (Guo, Cocks et al., 1997), scanning electron microscopy (Imahori, 2010), and X-ray photoelectron spectroscopy (Westermark, Rensmo et al., 2002), however each of these techniques suffers from the requirement for high vacuum. Our group recently investigated employing X-ray reflectivity under ambient conditions to convert the measured interference spectra (Figure 9b) into a dye thickness and subsequently a molecular orientation for a dye/TiO_2 bilayer (Wagner, Griffith et al., 2011). However, this technique is still limited by the need for a flat surface rather than measuring nanoporous DSSC electrodes directly. Despite experimental difficulties with confirming orientation, the design of porphyrin dyes which can inject both directly through space or facilitated by the linker group presents a promising method for enhancing overall injection.

In addition to modifying the dye structure to enhance injection efficiency, there are a range of additives which can be introduced to the electrolyte or sensitizing dye bath solutions to achieve enhanced injection. For instance, one potential issue with injection in porphyrin-sensitized solar cells is the limited free energy driving forces available for some dyes. This becomes a problem for dyes with a large red-shift in the standard porphyrin absorption spectrum, and in particular, the free base porphyrin dyes, which can often display LUMO energies approaching that of the semiconductor conduction band potential. The absence of significant free energy driving forces is intrinsic to the dye/semiconductor combination, and is difficult to alter with structural modifications of the dye. However, the conduction band edge potential (E_{CB}) is related to the surface potential of the oxide. Introducing charged species into the electrolyte which subsequently adsorb to the semiconductor surface can therefore shift the value of E_{CB} and change the relative driving force for injection. Placing alkali metal cations in the electrolyte is the most common way to achieve a positive shift of E_{CB}, thereby improving the injection driving force for dyes with low (more positive) LUMO energies (Liu, Hagfeldt et al., 1998). Another additive which has been shown to improve injection in porphyrin-sensitized solar cells is chenodeoxycholic acid (CDCA). This additive is generally dissolved in the sensitizing dye solution and acts to prevent aggregation of the dyes on the surface, a significant issue for porphyrin sensitizers, which interact strongly through π-π stacking forces (Planells, Forneli et al., 2008). Surface aggregation induces injection from excited dyes into neighbouring dye molecules, thus reducing the injection efficiency through a self-quenching mechanism. CDCA molecules co-adsorb to the oxide surface with the dye, preventing aggregate formation and elevating the injection efficiency.

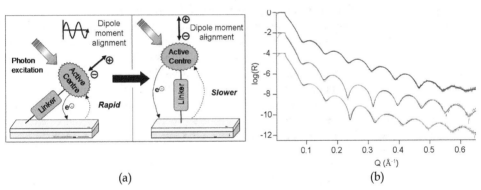

(a)   (b)

Fig. 9. (a) An illustration of the effect of dye adsorption orientation on the charge transfer and dipole alignment at a dye sensitised electrode. (b) Observed (data points) and calculated (solid lines) X-ray reflectivity spectra for a TiO_2 substrate (red), and porphyrin-sensitized TiO_2 before (blue) and after (green) 1 hour light exposure. Figure 9b taken from (Wagner, Griffith et al., 2011) and reproduced by permission of The American Chemical Society.

An alternative method to electrolyte additives which can be employed to modulate the semiconductor conduction band is to change the material employed as the semiconductor. The density of states (DOS) distribution for semiconductors is normally expressed as an exponential function with a characteristic broadening parameter, unique for each different metal oxide. As such, different materials will display various potentials at matched electron densities, leading to different E_{CB} values (Grätzel, 2001). In order to obtain a more positive E_{CB} to enhance the driving force for injection, the standard TiO_2 semiconductor can be replaced with materials such as SnO_2 (Fukai, Kondo et al., 2007), In_2O_3 (Mori & Asano, 2010) or WO_3 (Zheng, Tachibana et al., 2010), which all possess a narrower DOS distribution and thus lower E_{CB} values than TiO_2 at the same charge densities. Each of these materials produce higher photocurrents than TiO₂-based systems due to enhanced injection, however the electron mobility in these oxides is much higher than in TiO₂ and thus they suffer from faster recombination reactions which minimize or can even reverse the overall efficiency gains achieved by enhancing injection.

The injection yield of porphyrin-sensitized devices can also be improved by innovative device design or the use of various post-treatments to improve the system. Our group recently explored such post-treatments, demonstrating improvements in the J_{sc} of a zinc porphyrin DSSC arsing from enhanced injection after the cell was exposed to AM 1.5 illumination for 1 hr (Wagner, Griffith et al., 2011). The injection yield was measured using absorbed photon-to-current conversion efficiency (APCE), which is calculated by normalizing the IPCE for light absorption:

$$APCE = \frac{IPCE}{LHE} = \phi_{inj}\, \eta_{coll} \qquad (7)$$

By employing thin (~2 µm) film DSSCs, transport losses are assumed to be negligible and thus η_{coll} is close to 100% and the APCE measurements enable determination of ϕ_{inj} under short circuit conditions. The increased APCE (from 65% to approximately 90%) following light exposure (Figure 10a) therefore demonstrated an increased injection yield for the porphyrin dye. We have also employed APCE measurements to demonstrate an enhancement in the injection yield when zinc and free base porphyrin dyes were combined on the same TiO₂ surface. The APCE of the mixture was ~300% higher than either individual dye. It was proposed that this enhanced injection could arise from energy transfer from the zinc dye with an inefficient linker to the free base dye which possesses a conjugated linker, possible due to the spectral overlap between zinc porphyrin emissions and free base porphyrin absorption (Griffith, Mozer et al., 2011). This process could allow the zinc dye to inject through a more efficient conjugated pathway on the free base dye (Figure 10b).

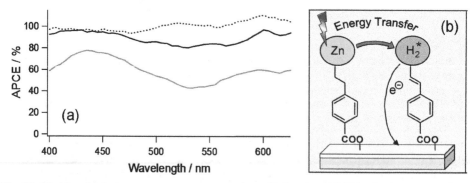

Fig. 10. (a) Absorbed photon to current conversion efficiencies (APCE) which estimate the injection yield for porphyrin-sensitized thin-film TiO₂ devices before (grey solid line) and after (black solid line) 1 hour light exposure. Data for the N719 dye is included for comparison (dashed line). (b) Energy transfer from a zinc to a free base porphyrin to utilize the conjugated injection pathway. Figure 10a taken from (Wagner, Griffith et al., 2011) and reproduced by permission of The American Chemical Society.

2.3 Charge transport

Since the nanoparticles of typical DSSC anodes are too small to sustain a space charge layer, electron transport in DSSCs is dominated by diffusion with negligible drift contributions. In this situation, the charge collection efficiency, η_{coll}, is related to the electron diffusion coefficient (D) and electron lifetime (τ) in the semiconductor electrode (where electron lifetime is the average time spent in the electrode). If the electron diffusion length, L, where:

$$L = \sqrt{D\tau} \qquad (8)$$

is shorter than the thickness of the semiconductor electrode, then electrons will recombine with the dye cation or the acceptor species in the redox mediator during charge transport, limiting η_{coll}. Typical diffusion lengths for the benchmark ruthenium dyes are 30-60 μm, leading to high collection efficiencies on 20 μm semiconductor films. The diffusion coefficients for porphyrin DSSCs are comparable to most other dyes. However, many porphyrins, and in particular free base dyes, suffer from high levels of recombination which lower the electron lifetime and thus the diffusion length. The effective diffusion length of sensitizers can be estimated from the film thickness at which the measured IPCE or J_{sc} saturates. However, such measurements cannot deconvolute the competing affects of increasing light harvesting and decreasing collection efficiency. Since the film thickness required for unity absorption of incident photons is ~6 μm, J_{sc} saturation values below this limit suggest there will be charge transport losses, as has been measured for some porphyrin DSSCs (Figure 11a). To determine L, D and τ values more rigorously, small amplitude perturbation techniques such as intensity modulated photovoltage or photocurrent spectroscopy, impedance spectroscopy or stepped-light induced measurements of photocurrent and photovoltage are generally employed, producing plots such as the one displayed in Figure 11b. However, there is some debate regarding the accuracy of these transient techniques, with Barnes et al. arguing that IPCE measurements performed with front and backside illumination are more relevant than small perturbation relaxation techniques (Barnes, Liu et al., 2009). In order to remove or minimize the charge transport losses in some porphyrins, strategies which reduce the recombination must be explored.

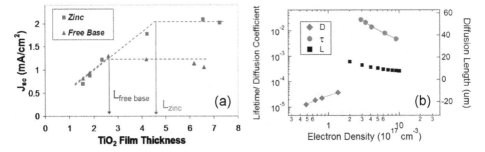

Fig. 11. (a) Diffusion length estimated from J_{sc} saturation values for inefficient zinc and free base porphyrins. (b) D (blue diamonds) and τ (red circles) values measured by stepped light-induced photovoltage and photocurrent techniques plotted against electron density for a porphyrin-sensitized DSSC. The calculated electron diffusion length, L, is also shown (black squares).

3. Charge recombination

As described earlier, the J_{sc} of porphyrin-sensitized solar cells is determined by their spectral response, injection efficiency and charge transport characteristics, all of which are quite well understood. Conversely, the open circuit voltage (V_{oc}) of porphyrin DSSCs is generally observed to be 100–200 mV lower than the commonly used ruthenium dyes, the origin of which is only partially elucidated. Since the photovoltage under illumination is dependent on the Fermi level in the semiconducting oxide, the lower V_{oc} for porphyrin DSSCs may be related to either a positive shift of the conduction band potential (E_{CB}) of the semiconducting oxide following dye sensitization or a lower electron density due to a reduced electron lifetime. Our group investigated each of these possibilities in collaboration with Japanese co-workers in order to determine the origin of the lower V_{oc} in porphyrin DSSCs. It was found that when the V_{oc} was plotted against the electron density (ED) in the TiO$_2$ film, neither the slope nor the y-intercept of the V_{oc} vs logED plot differed between ruthenium and porphyrin sensitized solar cells (Mozer, Wagner et al., 2008) (Figure 12d). Since the redox mediator Fermi level was constant in each case, the V_{oc} vs logED plot is indicative of the TiO$_2$ conduction band potential. Hence these results demonstrated that the lower V_{oc} of porphyrin-sensitized solar cells is not due to an E_{CB} shift following dye uptake. We found instead that the low photovoltages were a result of electron lifetimes in porphyrin dyes being reduced by a factor of ~200 at matched electron densities, independent of their chemical structure (Figure 12b). Furthermore, we showed that the shorter electron lifetimes were not related to electron transport differences, since the diffusion coefficients were identical for porphyrin and ruthenium dyes (Figure 12c).

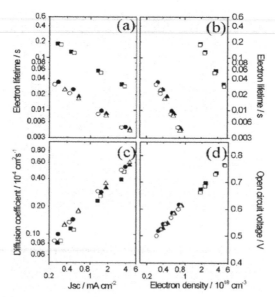

Fig. 12. (a) Electron lifetime and (c) diffusion coefficient versus short circuit current density. (b) Electron lifetime and (d) open circuit voltage versus electron density for ruthenium (squares) and porphyrin (circles, triangles) DSSCs. Figure taken from (Mozer, Wagner et al., 2008) and reproduced by permission of The Royal Society of Chemistry

Since charge is a conserved quantity in any system, a continuity equation for the charge density, n, can be derived for a DSSC. The time-dependent form of this equation is:

$$\frac{\partial n}{\partial t} = \phi_{inj}\, \alpha\, I_0\, \exp(-\alpha x) + D_0 \frac{\partial^2 n}{\partial x^2} - \left(\frac{n}{\tau_{redox}} + \frac{n}{\tau_{dye}} \right) \tag{9}$$

where the first term on the right-hand side of the equation describes the electron injection into the oxide from dyes at position x (α is the absorption coefficient, I_0 is the incident photon flux and $x = 0$ at the anodic contact). The second term accounts for the diffusion of electrons (D_0 is the diffusion coefficient of electrons), whilst the third term describes the two simultaneously occurring recombination reactions (where τ_{redox} and τ_{dye} are the lifetimes determined by the recombination reactions of conduction band electrons with the redox acceptor species and the oxidised dye, respectively). Since the lower V_{oc} of porphyrin DSSCs arises from a reduced electron lifetime which is not affected by electron transport, it must be related to an enhancement in one (or both) of the two recombination processes.

Dye cation recombination in DSSCs has been extensively studied using transient absorption spectroscopy to probe the rate of disappearance of the dye cation absorption following its creation. For the majority of dyes, the cations are regenerated with a time constant of 1-10 µs, even in viscous or semi-solid electrolytes which slow down the reaction due to diffusion limitations (Nogueira & Paoli, 2001; Wang, Zakeeruddin et al., 2003). These kinetics are generally much faster than the recombination reaction between dye cations and electrons in the semiconductor, which has a time constant of 100 µs – 1 ms (Willis, Olson et al., 2002). Our group has demonstrated this situation holds true for porphyrin dyes by measuring transient absorption kinetics for the dye cation (with an absorption peak at 700 nm) in the absence and the presence of a standard I^-/I_3^- redox mediator (Figure 13). Without the redox mediator the half signal decay was 60 µs, whilst in the presence of the redox mediator, the half-signal decay was accelerated to 2 µs (Wagner, Griffith et al., 2011). This suggests efficient prevention of recombination through regeneration of the dye cations by the redox mediator. It is therefore very unlikely that the short electron lifetime for porphyrin DSSCs results from recombination with the dye cation.

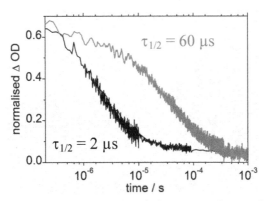

Fig. 13. Transient absorption kinetic traces recorded at 700 nm for porphyrin-sensitized TiO$_2$ films covered with acetonitrile electrolyte in the absence (red) and presence (black) of an I$^-$/I$_3^-$ redox mediator. The films were photoexcited by nanosecond pulses at 532 nm.

As dye cation recombination is a negligible problem for porphyrin DSSCs, the shorter electron lifetime must arise from increased recombination between conduction band electrons and the acceptor species in the redox mediator. Such a process can only occur from an increased proximity of the acceptor species to the semiconductor surface. For the standard I^-/I_3^- redox mediator, it has been proposed that most organic dyes (specifically including porphyrins) either attract I_3^- to the dye–semiconductor interface (Miyashita, Sunahara et al., 2008) or catalyse the recombination reaction with acceptor species in the electrolyte, such as I_3^- or the iodine radical I_2^- (O'Regan, López-Duarte et al., 2008). Several different strategies have been implemented in an attempt to improve the electron lifetime, and we now examine some of the major innovations which have lead to enhancements in the overall device V_{oc}.

3.1 Molecular structure

The molecular structure of dyes can have a large impact on the concentration of the redox mediator at the semiconductor surface. Nakade et al. reported that adsorption of ruthenium dye N719 will decrease the concentration of acceptor species I_3^- in the vicinity of the TiO_2 surface due to shielding from the negative SCN^- ligands on the dye molecule (Nakade, Kanzaki et al., 2005). A similar physical shielding effect can be achieved with organic dyes by introducing bulky substituent groups to sterically hinder the approach of the redox mediator to the semiconductor surface (Koumura, Wang et al., 2006) (Figure 14). This approach was shown to increase the electron lifetime and V_{oc} for DSSCs constructed with carbazole (Miyashita, Sunahara et al., 2008), phthalocyanine (Mori, Nagata et al., 2010) and osmium (Sauvé, Cass et al., 2000) complexes. Several of these authors reported minimal effects when the dye loading on the surface was reduced, confirming that the structure of the dye, and its steric crowding of the semiconductor surface, was the major factor driving the increase in electron lifetime. This strategy has been successfully implemented to porphyrin sensitizers, with the introduction of octyl chains to a high efficiency zinc porphyrin dye producing the highest efficiency ionic liquid-based porphyrin DSSC (Armel, Pringle et al., 2010). Imahori et al. have demonstrated the value of amending the porphyrin structure by adding bulky mesityl groups at the meso positions of the porphyrin core to both reduce the dye aggregation (which limits electron injection) and enhance the V_{oc} by blocking the surface from the approach of the redox mediator.

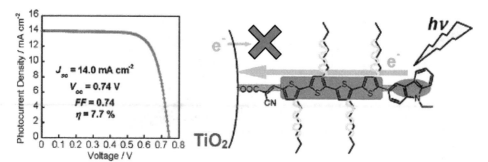

Fig. 14. PV parameters and an illustration of a carbazole dye with long alkyl chains to stop the redox mediator reaching the semiconductor surface. Figure taken from (Koumura, Wang et al., 2006) and reproduced by permission of The American Chemical Society.

3.2 Semiconductor blocking effects

As was earlier described for electron injection, strategies to inhibit recombination between conduction band electrons and the redox mediator can focus on either the dye or the semiconductor side of the major electronic interface. Accordingly, an alternative strategy to dye structure modification which can be employed to extend the electron lifetime in porphyrin DSSCs is to insulate exposed sites on the semiconductor surface. Rather than block the approach of the redox mediator to these active sites, this method attempts to deactivate the electron transfer process at these sites using an insulating surface covering. Deposition of a compact TiO_2 layer from a titanium tetrachloride ($TiCl_4$) precursor has been previously used to block electron transfer between the redox mediator and the back FTO-glass contact (Burke, Ito et al., 2008), and the same approach has also been successfully applied to insulating the semiconductor surface. O'Regan et al. utilized photocurrent and photovoltage transient measurements to show that deposition of a compact TiO_2 blocking layer on top of the mesoporous TiO_2 electrode produces an 80 mV downward shift in the TiO_2 conduction band edge potential and a 20-fold decrease in the electron/electrolyte recombination rate constant (O'Regan, Durrant et al., 2007). Following these findings, a range of organic acids have been trialled as surface insulating agents. Phosphinic acids are particularly useful in this regard since they form strong bonds with titania but, in contrast to commonly employed carboxylic or phosphonic acids, also have two organic substituents which can potentially provide more complete insulation of the semiconductor surface. Accordingly our group, in collaboration with Australian co-workers, employed a phosphinic acid surface treatment to a zinc porphyrin DSSC and demonstrated a successful suppression of the surface recombination and a simultaneous positive conduction band shift, resulting in 15% improvements in the photocurrent and 20% increases in the overall device efficiency. Measurements of time-resolved photovoltage transients demonstrated that these improvements resulted from an increased electron lifetime (Figure 15a), although the expected V_{oc} improvement was limited by a simultaneous positive shift in the semiconductor conduction band potential (Allegrucci, Lewcenko et al., 2009) (Figure 15b). Nonetheless, these results establish that the short electron lifetimes which limit porphyrin DSSCs can be improved with a semiconductor surface treatment.

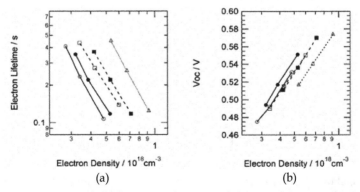

Fig. 15. (a) Electron lifetime, and (b) V_{oc}, as a function of electron density in the TiO_2 film for porphyrin DSSCs after 0 mins (circles), 5 mins (squares) and 30 mins (triangles) of a phosphinic acid surface treatment. Figure taken from (Allegrucci, Lewcenko et al., 2009) and reproduced by permission of The Royal Society of Chemistry.

3.3 Manipulating interfacial charges

The predominate recombination pathway in porphyrin DSSCs is between electrons and the acceptor species in the redox mediator. Consequently, the cations and additives which are typically dissolved in the electrolyte play an important role in mediating this reaction. The roles of the cations have been found to influence the electron injection yield, the open-circuit voltage, the electron diffusion coefficient, and the rate of dye-cation regeneration (Kambe, Nakade et al., 2002; Zaban, Ferrere et al., 1998). With careful design, the influence of these supporting cations can be manipulated to remove the acceptor species in the redox mediator from the vicinity of the semiconductor surface, thereby extending the electron lifetime and raising the V_{oc}. Nakade et al investigated such effects by varying the size of the cation additive for a ruthenium-sensitized solar cell with a standard I^-/I_3^- redox mediator. They found that the size of the cation has a large impact on the thickness of the electrical double layer (Helmholtz and diffuse layers), effectively altering the local (surface) concentration of I_3^-, which is the concentration of I_3^- within the distance from the TiO_2 surface at which electrons can be transferred (Nakade, Kanzaki et al., 2005). When electrons are injected into the TiO_2, the surface becomes negatively charged and an electrical double layer is formed at the surface. For cations which are small enough to penetrate between the adsorbed dye molecules this double layer is formed over ~1 nm, effectively screening the surface charge and allowing I_3^- to approach close to the TiO_2 surface. However, for bulky cations such as tetrabutylammonium (TBA$^+$) which cannot penetrate between the dye and TiO_2, a distance much longer than the size of the dye is needed for the screening. In this case, anions feel a repulsive force to penetrate between the dye and TiO_2 due to the negative surface charge. This reduces the local I_3^- concentration and results in a longer electron lifetime (Figure 16).

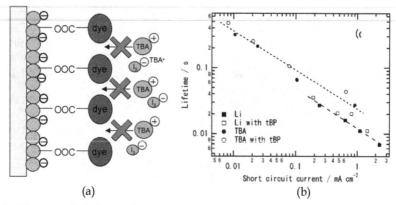

(a)   (b)

Fig. 16. (a) The extended electric double layer at the surface using a bulky supporting cation. (b) The electron lifetime for DSSCs employing TBA$^+$ and Li$^+$ in the electrolyte. Figure 16b taken from (Nakade, Kanzaki et al., 2005) and reproduced by permission of The American Chemical Society.

Nakade et al. showed that this increased electron lifetime led to a 300 mV increase in V_{oc}, although the increase was aided by a negative shift in E_{CB} which reduced the photocurrent. Since such an approach simply manipulates the position of intrinsically formed charged layers using a generic electrolyte, it should be generally applicable to all dye systems, including porphyrins.

Post-treatments or dye interactions could also lead to longer electron lifetimes and improved V_{oc} in porphyrin DSSCs. Our group have demonstrated using photovoltage decay measurements that exposure of a zinc porphyrin DSSC to AM 1.5 illumination conditions for a period of 1 hour produces an increase in the electron lifetime by a factor of 2 to 3. This result was also accompanied by a comparable decrease in the electron diffusion coefficient. The improved electron lifetime combined with the increased J_{sc} obtained from the same post-treamtment resulted in increased electron densities at open circuit conditions, leading to improved V_{oc} (Wagner, Griffith et al., 2011). It was postulated that the origin of this effect could be either the photo-generation of electronic states within the band gap of TiO_2 or a change in the behavior of electrolyte addtives when the solar cell is illmuinated, both of which could lead to improved injection, longer electron lifetimes and slower electron transport.

Charge transfer interactions could also act to decrease recombination in porphyrin DSSCs in certain circumstances. Our group recently reported an enhanced injection yield when zinc and free base porphyrin dyes were combined, however we also noted that this mixture resulted in a higher V_{oc} than that obtained from both individual dyes. Measured energy levels for the two dyes indicate that the zinc dye (ZnNC) had both a higher HOMO and a higher LUMO energy than the free base dye (FbC), which could lead to hole transfer from FbC$^+$ to neutral ZnNC (Figure 17a). It was noted that similar charge transfer processes between zinc and free base porphyrins have been previously observed to occur on very fast (picosecond) timescales (Koehorst, Boschloo et al., 2000). It was speculated that hole transfer (HT) could potentially improve the charge generation yield of FbC by preventing recombination. This would be feasible if $k_{HT} \gg k_{EDR,FbC} \gg k_{DR,FbC}$, where $k_{EDR,FbC}$ is the rate constant for charge recombination between TiO_2 electrons and FbC$^+$, and $k_{DR,FbC}$ is the rate constant for dye regeneration of FbC (Griffith, Mozer et al., 2011) (Figure 17b). Such charge transfer processes have been shown to reduce recombination and improve the V_{oc} for other co-sensitized DSSC systems (Clifford, Forneli et al., 2011; Clifford, Palomares et al., 2004), and offer an attractive pathway to simultaneously remove both injection and recombination limitiations in porphyrin DSSCs.

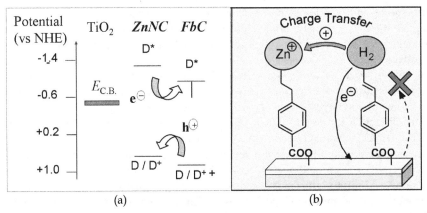

Fig. 17. (a) Calculated dye HOMO and LUMO energy levels and potential charge transfer processes for a zinc/free base mixed dye system. (b) Charge transfer from oxidized free base to neutral zinc molecules to prevent recombination.

4. Conclusion

The efficient light harvesting potential of porphyrin dyes, exemplified by their primary role in photosynthesis, makes them ideal candidates for use as photosensitizers within dye sensitized solar cells. Their synthesis is relatively straightforward, and their optical and electronic properties can be tuned via chemical modification of the coordinating metal centre, the porphyrin core, the number of porphyrin units, and the linker between the core and the inorganic oxide. Recent porphyrin DSSC developments have been accompanied by a simultaneous improvement in the understanding of the photophysics governing operational solar cells. In particular, many of the fundamental limitations which constrain the performance of these dyes have been elucidated.

The major limitations which continue to hinder the performance of porphyrin DSSCs are the light harvesting of incident photons, injection into the semiconducting oxide and the recombination with the acceptor species in the redox mediator. Light harvesting limitations, which mainly surround the limited absorption of low energy (red) photons, can be circumvented by combining several dyes with complimentary absorption spectra or by employing multichromophore dyes to boost the effective absorption coefficients and allow thinner semiconductor films to be employed. Electron injection yields for porphyrin dyes, studied by techniques including time resolved luminescence quenching, ultrafast transient absorption spectroscopy and absorbed photon-to-current conversion efficiency, have been shown to be much lower than the kinetics of injection compared to dye deactivation would predict. Such limitations can be caused by dye structural considerations, heterogeneous injection kinetics, or poor free energy driving forces. These limitations can be addressed by modifying the dye structure, addition of various chemicals to the electrolyte to modify the free energy driving force for injection, employing post-treatments to enhance injection efficiency after fabrication or by combining different dyes to achieve improved injection efficiencies through synergistic interactions. Recombination limitations, now understood to be the major impediment to achieving high efficiency in porphyrin DSSCs, have been shown to arise from the back reaction between electrons in the oxide conduction band and the acceptor species in the redox mediator. Characterization of the electron lifetime, studied by techniques such as intensity modulated voltage spectroscopy (IMVS), electrochemical impedance spectroscopy and stepped-light induced measurements of photocurrent and photovoltage (SLIM-PCV) reveals that this recombination reaction can be influenced by several factors, such as a physical blocking effect on either the dye structure or the semiconductor surface, electrostatic interactions which control the location of charges at the interface or combining dyes to harness various photoinduced charge transfer mechanisms.

Given the attractive features of porphyrin chromophores, the improved understanding of porphyrin DSSCs which has been compiled in recent years, and the many innovative strategies still emerging, there remains much promise in the development of these devices. Porphyrin based DSSCs continue to offer a fruitful topic for exploring the fundamental processes which limit the efficiency of dye-sensitized light harvesting applications, inspiring the development of innovative strategies to circumvent these basic limitations. The remaining challenge is to integrate each of these new strategies to produce a porphyrin DSSC with a power conversion efficiency which surpasses the current maximum of 11% and allows these devices to become a commercial reality.

5. Acknowledgments

The authors gratefully acknowledge the financial support of the Australian Research Council through the ARC Centre of Excellence Federation Fellowship, Discovery, and LIEF schemes. MJG acknowledges the additional support of an Australian Postgraduate Award and a Prime Minister's Asia Australia Endeavour Award from the Australian Federal government. The authors would like to thank, in no particular order, Prof. David Officer, Prof. Gordon Wallace, Dr Kaludia Wagner, Dr Pawel Wagner, Prof. Keith Gordon, Dr Ryuzi Katoh, Assoc. Prof. Akihiro Furube and Assoc. Prof. Shogo Mori for their invaluable collaborations and fruitful discussions.

6. References

Allegrucci, A., Lewcenko, N. A., et al. (2009). "Improved performance of porphyrin-based dye sensitised solar cells by phosphinic acid surface treatment." *Energy and Environmental Science*, Vol. 2, pp 1069.

Armel, V., Pringle, J. M., et al. (2010). "Ionic liquid electrolyte porphyrin dye sensitised solar cells." *Chemical Communications*, Vol. 46, pp 3146.

Bach, U., Lupo, D., et al. (1998). "Solid-state dye-sensitized mesoporous TiO_2 solar cells with high photon-to-electron conversion efficiencies." *Nature*, Vol. 395, pp 583.

Barnes, P. R., Liu, L., et al. (2009). "Re-evaluation of recombination losses in dye-sensitized cells: the failure of dynamic relaxation methods to correctly predict diffusion length in nanoporous photoelectrodes." *Nano Letters*, Vol. 9, pp 3532.

Bessho, T., Zakeeruddin, S. M., et al. (2010). "Highly efficient mesoscopic dye-sensitized solar cells based on donor–acceptor-substituted porphyrins." *Angewandte Chemie (International Edition)*, Vol. 49, pp 6646.

Bisquert, J. & Mora-Sero, I. (2010). "Simulation of steady-state characteristics of dye-sensitized solar cells and the interpretation of the diffusion length." *Journal of Physical Chemistry Letters*, Vol. 1, pp 450.

Boschloo, G. & Hagfeldt, A. (2009). "Characteristics of the iodide/triiodide redox mediator in dye-sensitized solar cells." *Accounts of Chemical Research*, Vol. 42,(11), pp 1819.

Burke, A., Ito, S., et al. (2008). "The function of a TiO_2 compact layer in dye-sensitized solar cells incorporating "planar" organic dyes." *Nano Letters*, Vol. 8, pp 977.

Campbell, W. M., Burrell, A. K., et al. (2004). "Porphyrins as light harvesters in the dye sensitized solar cell." *Coordination Chemistry Reviews*, Vol. 248, pp 1363.

Campbell, W. M., Jolley, K. W., et al. (2007). "Zn porphyryrins as highly efficient sensitizers in dye sensitized solar cells." *Journal of Physical Chemistry C*, Vol. 111, pp 11760.

Chiba, Y., Islam, A., et al. (2006). "Dye sensitized solar cells with conversion efficiency of 11.1%." *Japanese Journal of Applied Physics*, Vol. 48, pp L638.

Clifford, J. N., Forneli, A., et al. (2011). "Co-sensitized DSCs: dye selection criteria for optimized device V_{oc} and efficiency " *Journal of Materials Chemistry*, Vol. 21, pp 1693.

Clifford, J. N., Palomares, E., et al. (2004). "Multistep electron transfer processes on dye co-sensitized nanocrystallineTiO_2 films." *Journal of the American Chemical Society*, Vol. 126, pp 5670.

Clifford, J. N., Yahioglu, G., et al. (2002). "Molecular control of recombination dynamics in dye sensitised nanocrystalline TiO_2 films." *Chemical Communications*, Vol., pp 1260.

Dos Santos, T., Morandeira, A., et al. (2010). "Injection limitations in a series of porphyrin dye sensitized solar cells." *Journal of Physical Chemistry C*, Vol. 114, pp 3276.

Fukai, Y., Kondo, Y., et al. (2007). "Highly efficient dye-sensitized SnO_2 solar cells having sufficient electron diffusion length " *Electrochemistry Communications*, Vol. 9, pp 1439.

Furube, A., Katoh, R., et al. (2005). "Lithium ion effect on electron injection from a photoexcited coumarin derivative into a TiO_2 nanocrystalline film investigated by visible-to-IR ultrafast spectroscopy." *Journal of Physical Chemistry B*, Vol. 109, pp 16406.

Geng, L. & Murray, R. W. (1986). "Oxidative microelectrode voltammetry of tetraphenylporphyrin and copper tetraphenylporphyrin in toluene solvent." *Inorganic Chemistry*, Vol. 25, pp 3115.

Gledhill, S. E., Scott, B., et al. (2005). "Organic and nano-structured composite photovoltaics: An overview " *Journal Material Research*, Vol. 20, pp 3167.

Gouterman, M. (1978). "Electronic Spectra", In: *The Porphyrins (Vol. III)*. D. Dolphin (ed.), Academic Pess Inc., New York.

Grätzel, M. (2001). "Photoelectrochemical cells." *Nature*, Vol. 414, pp 338.

Grätzel, M. (2005). "Solar energy conversion by dye sensitized photovoltaic cells." *Inorganic Chemistry*, Vol. 44, pp 6841.

Griffith, M. J., Mozer, A. J., et al. (2011). "Remarkable synergistic effects in a mixed porphyrin dye-sensitized TiO_2 film." *Applied Physics Letters*, Vol. 98, pp 163502.

Guo, Q., Cocks, I., et al. (1997). "The orientation of acetate on a $TiO_2(110)$ surface." *Journal of Physical Chemistry*, Vol. 106,(7), pp 2924.

Haque, S. A., Palomares, E., et al. (2005). "Charge separation versus recombination in dye sensitized solar cells: The minimization of kinetic redundancy." *Journal of the American Chemical Society*, Vol. 127, pp 3456.

Hara, K., Sato, T., et al. (2003). "Molecular design of coumarin dyes for efficient dye sensitized solar cells." *Journal of Physical Chemistry B*, Vol. 107, pp 597.

Hardin, B. E., Hoke, E. T., et al. (2009). "Increased light harvesting in dye-sensitized solar cells with energy relay dyes." *Nature Photonics*, Vol. 3, pp 406.

Hengerer, R., Kavan, L., et al. (2000). "Orientation dependence of charge-transfer processes on TiO_2 (anatase) single crystals." *Journal of the Electrochemical Society*, Vol. 147,(4), pp 1467.

Hsieh, C.-P., Lu, H.-P., et al. (2010). "Synthesis and characterization of porphyrin sensitizers with various electron-donating substituents for highly efficient dye-sensitized solar cells." *Journal of Materials Chemistry*, Vol. 20, pp 1127.

Imahori, H. (2010). "Porphyrins as potential sensitizers for dye sensitized solar cells." *Key Engineering Materials*, Vol. 451, pp 29.

Imahori, H., Umeyama, T., et al. (2009). "Large π-aromatic molecules as potential sensitizers for highly efficient dye sensitized solar cells." *Accounts of Chemical Research*, Vol. 42, pp 1809.

Kambe, S., Nakade, S., et al. (2002). "Influence of the electrolytes on electron transport in mesoporous TiO_2 electrolyte systems." *Journal of Physical Chemistry B*, Vol. 106, pp 2967.

Katoh, R., Furube, A., et al. (2002). "Efficiencies of electron injection from excited sensitizer dyes to nanocrystalline ZnO films as studied by near-IR optical absorption of injected electrons." *Journal of Physical Chemistry B*, Vol. 106, pp 12957.

Kay, A. & Grätzel, M. (1993). "Artificial photosynthesis. 1. Photosensitization of TiO_2 solar cells with chlorophyll derivatives and related natural porphyrins." *Journal of Physical Chemistry*, Vol. 97, pp 6272.

Koehorst, R. B. M., Boschloo, G. K., et al. (2000). "Spectral sensitization of TiO_2 substrates by monolayers of porphyrin heterodimers." *Journal of Physical Chemistry B*, Vol. 104, pp 2371.

Koops, S. E. & Durrant, J. R. (2008). "Transient emission studies of electron injection in dye sensitised solar cells." *Inorganica Chimica Acta*, Vol. 361, pp 8.

Koumura, N., Wang, Z.-S., et al. (2006). "Alkyl-functionalized organic dyes for efficient molecular photovoltaics." *Journal of the American Chemical Society*, Vol. 128, pp 14256.

Kuang, D., Ito, S., et al. (2006). "Stable mesoscopic dye-sensitized solar cells based on tetracyanoborate ionic liquid electrolyte." *Journal of the American Chemical Society*, Vol. 128, pp 4146.

Lee, C.-W., Lu, H.-P., et al. (2009). "Novel zinc porphyrin sensitizers for dye sensitized solar cells: Synthesis and spectral, electrochemical, and photovoltaic properties." *Chemistry; A European Journal*, Vol. 15, pp 1403.

Liu, C.-Y., Tang, H., et al. (1996). "Effect of orientation of porphyrin single crystal slices on optoelectronic properties." *Journal of Physical Chemistry*, Vol. 100, pp 3587.

Liu, Y., Hagfeldt, A., et al. (1998). "Investigation of influence of redox species on the interfacial energetics of a dye-sensitized nanoporous TiO_2 solar cell " *Solar Energy Materials and Solar Cells*, Vol. 55, pp 267.

Lo, C.-F., Hsu, S.-J., et al. (2010). "Tuning spectral and electrochemical properties of porphyrin-sensitized solar cells." *Journal of Physical Chemistry C*, Vol. 114, pp 12018.

Marcus, R. A. (1964). "Chemical and electrochemical electron-transfer theory." *Annual Review of Physical Chemistry*, Vol. 15, pp 155.

Martinson, A. B. F., Elam, J. W., et al. (2007). "ZnO nanotube based dye sensitized solar cells." *Nano Letters*, Vol. 8, pp 2183.

Miyashita, M., Sunahara, K., et al. (2008). "Interfacial electron-transfer kinetics in metal-free organic dye sensitized solar cells: Combined effects of molecular structure of dyes and electrolytes." *Journal of the American Chemical Society*, Vol. 130, pp 17874.

Mor, G. K., Shankar, K., et al. (2006). "Use of highly-ordered TiO_2 nanotube arrays in dye-sensitized solar cells." *Nano Letters*, Vol. 2, pp 215.

Mori, S. & Asano, A. (2010). "Light intensity independent electron transport and slow charge recombination in dye-sensitized In_2O_3 solar cells: In contrast to the case of TiO_2." *Journal of Physical Chemistry C*, Vol. 114, pp 13113.

Mori, S., Nagata, M., et al. (2010). "Enhancement of incident photon-to-current conversion efficiency for phthalocyanine-sensitized solar cells by 3D molecular structuralization." *Journal of the American Chemical Society*, Vol. 132, pp 4054.

Mozer, A. J., Griffith, M. J., et al. (2009). "Zn-Zn porphyrin dimer-sensitised solar cells: Towards 3-D light harvesting." *Journal of the American Chemical Society*, Vol. 131,(43), pp 15621.

Mozer, A. J., Wagner, P., et al. (2008). "The origin of open circuit voltage of porphyrin-sensitised TiO_2 solar cells." *Chemical Communications*, Vol., pp 4741.

Nakade, S., Kanzaki, T., et al. (2005). "Role of electrolytes on charge recombination in dye-sensitized TiO_2 solar cell (1): The case of solar cells using the I^-/I_3^- redox couple." *Journal of Physical Chemistry B*, Vol. 109, pp 3480.

Nazeeruddin, M. K., Pechy, P., et al. (2001). "Engineering of efficient panchromatic sensitizers for nanocrystalline TiO_2-based solar cells." *Journal of the American Chemical Society*, Vol. 123, pp 1613.

Nogueira, A. F. & Paoli, M.-A. D. (2001). "Electron transfer dynamics in dye sensitized nanoctystalline solar cells using a polymer electrolyte." *Journal of Physical Chemistry B*, Vol. 105, pp 7517.

O'Regan, B. & Grätzel, M. (1991). "A low-cost, high-efficiency solar cell based on dye-sensitized colloidal TiO_2 films." *Nature*, Vol. 353, pp 737.

O'Regan, B. C., Durrant, J. R., et al. (2007). "Influence of the $TiCl_4$ treatment on nanocrystalline TiO_2 films in dye-sensitized solar cells. 2. Charge density, band edge shifts, and quantification of recombination losses at short circuit." *Journal of Physical Chemistry C*, Vol. 111, pp 14001.

O'Regan, B. C., López-Duarte, I., et al. (2008). "Catalysis of recombination and its limitation on open circuit voltage for dye sensitized photovoltaic cells using phthalocyanine dyes." *Journal of the American Chemical Society*, Vol. 130, pp 2906.

Planells, M., Forneli, A., et al. (2008). "The effect of molecular aggregates over the interfacial charge transfer processes on dye sensitized solar cells." *Applied Physics Letters*, Vol. 92, pp 153506.

Sauvé, G., Cass, M. E., et al. (2000). "Dye sensitization of nanocrystalline titanium dioxide with osmium and ruthenium polypyridyl complexes." *Journal of Physical Chemistry B*, Vol. 104, pp 6821.

Shaheen, S. E., Ginley, D. S., et al. (2005). "Organic Based Photovoltaics: Towards Lower Cost Power Generation." *MRS Bulletin*, Vol. 20, pp 10.

Shockley, W. & Queisser, H. J. (1961). "Detailed balance limit of efficiency of p-n junction solar cells." *Journal of Applied Physics*, Vol. 32, pp 510.

Wagner, K., Griffith, M. J., et al. (2011). "Significant performance improvement of porphyrin-sensitised TiO_2 solar cells under white light illumination." *Journal of Physical Chemistry C*, Vol. 115, pp 317.

Wang, P., Zakeeruddin, S. M., et al. (2003). "A new ionic liquid electrolyte enhances the conversion efficiency of dye sensitized solar cells." *Journal of Physical Chemistry B*, Vol. 107, pp 13280.

Westermark, K., Rensmo, H., et al. (2002). "PES studies of Ru(dcbpyH2)2(NCS)2 adsorption on nanostructured ZnO for solar cell applications." *Journal of Physical Chemistry B*, Vol. 106, pp 10102.

Willis, R. L., Olson, C., et al. (2002). "Electron dynamics in nanocrystalline ZnO and TiO_2 films probed by potential step chronoamperometry and transient absorption spectroscopy." *Journal of Physical Chemistry B*, Vol. 106, pp 7605.

Zaban, A., Ferrere, S., et al. (1998). "Relative energetics at the semiconductor/sensitizing dye/electrolyte interface." *Journal of Physical Chemistry B*, Vol. 102, pp 452.

Zheng, H., Tachibana, Y., et al. (2010). "Dye sensitized solar cells based on WO_3." *Langmuir*, Vol. 26, pp 19148.

6

The Chemistry and Physics of Dye-Sensitized Solar Cells

William A. Vallejo L., Cesar A. Quiñones S. and Johann A. Hernandez S.
Universidad Nacional de Colombia,
Universidad de Cartagena
Universidad Distrital F.J.D.C, Bogotá,
Colombia

1. Introduction

Climate change is one of the major environmental problems that affect our society. At present annually more than 40 billons Tons of greenhouses gases are exhausted to atmosphere and the tendency is to the rise; the main reason for this situation is the high and uncontrolled use of fossil resource in energy generation. Development an environmental, friendly and reliable energy technology is a necessity. *Solar Energy* emerged as possible solution to confront this problem. This technology permits a direct conversion of sunlight into electrical power without exhaust of both greenhouse gases and another polluting agent. Actually silicon technology is market leader in photovoltaic technologies, however since a pioneering (Grätzel & O'Regan, 1991) dye-sensitized solar cells (DSSCs) have become in one important and promising technology in photovoltaic field. DSSCs given born to new solar cells generation replaced classical solid-state homo and hetero-junction device by a new concept with a nano-working electrode in photo-electrochemical cell. This technology offers a very low cost fabrication and easy industry introduction prospective; furthermore efficiencies near to 10% AM1.5 for DSSCs have been confirmed. DSSCs consists of three main components: A dye-covered nanoporous TiO_2 layer on a glass substrate coated with a transparent conductive oxide (TCO) layer, an redox electrolyte and a electrical contact deposited on conducting glass. Different parameters affect efficiency of the DSSCs: types of materials used as electrolyte, dye and electric contact, and synthesis method used to obtain these materials. In this chapter DSSCs components and different aspects related with photovoltaic principles and DSSCs performance will be studied. Special emphasis will put on to review physical, chemical and electrochemical principles of DSSCs operation.

2. Mechanism operation

All photovoltaic devices present two important steps to convert sunlight into electrical energy:
1. Radiation absorption with electrical excitation.
2. Charge carriers separation.

The way which radiation is absorbed and carriers are separated are two of the main differences between DSSCs and classical p-n junction. Conventional photovoltaic principle relies on differences in work functions between the electrodes of the cell in which photo-generated carriers could be separated, an asymmetry through cell is necessary to obtain electrical power. In classical p-n junction of solid state device the separation of photo-generated carriers relies on separation through depletion region built at p-n interface materials (Neamen, 1997). A different process occurs in DSSCs. Figure 1 shows typical scheme for DSSCs. The working electrode of DSSCs is conventionally constituted by mesoporous network of TiO_2 nanocrystalline (5-15μm, thickness) covered with a dye monolayer (conventionally Ru complex); this working electrode is supported on conducting glass (transparent conducting oxide, TCO). Different materials as platinum, palladium and gold could be use as counter-electrode of the cell; finally the gap between the electrodes is typically filled with a molten salt which containing a redox couple (A/A⁻); this salt is a hole conductor. Most DSSCs studied so far employ redox couple as iodide/tri-iodide (I^-/I_3^-) couple as electrolyte because of its good stability and reversibility (Pooman & Mehra), however others hole conductors as solid and ionic electrolytes also can be used. In overall process, the DSSCs generate electric power from light without suffering any permanent chemical transformation (Kelly & Meyer, 2001)

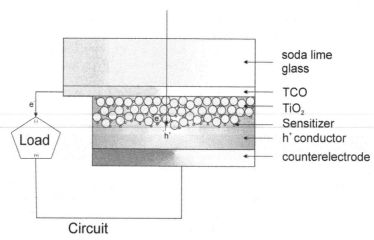

Fig. 1. General Structure of a dye-sensitized solar cell, the electron migration is showing.

In DSSCs, the basic photovoltaic principle relies on the visible photo-excitation of dye molecule; the esquematic reaction of overall process is follows:

$$TiO_2/D + h\nu \leftrightarrow TiO_2/D^*_{LUMO} \tag{1}$$

$$D^*_{LUMO} + CB_{TiO2} \rightarrow TiO_2/D^+ + e_{CB}^- \tag{2}$$

$$Pt + [I_3]^- \rightarrow 3I^- \tag{3}$$

$$TiO_2/D^+ + 3I^- \rightarrow [I_3]^- + D \tag{4}$$

$$e_{CB}^- + D_{HOMO} \rightarrow D + CB_{TiO2} \tag{5}$$

$$e_{CB}^- + [I_3]^- \rightarrow 3I^- + CB_{TiO2} \tag{6}$$

Where CB_{TiO2} is TiO_2 conduction band and D is dye molecule. First, an electron is photoexcited from highest occupied molecular orbital (HOMO) level to lowest unoccupied molecular orbital (LUMO) level into dye molecule (eq. 1). Then, electron injection from excited dye (D*) into TiO_2 conduction band occurs (eq. 2), excitation of dye usually is a transition-metal complex whose molecular properties are specifically for the task is able to transfer an electron to TiO_2 by the injection process. After that, electron migrates through TiO_2 network toward the TCO susbtrate (fig. 2(a)). The physics of charge transfer and transport in molecular and organic materials is dominated by charge localization resulting from polarization of the medium and relaxation of molecular ions. As a result of weak intermolecular interactions, the carriers in these materials are strongly localized on a molecule, and transport occurs via a sequence of charge-transfer steps from one molecule to other, similar to the hopping between defects states in inorganic semiconductors or band gap states in amorphous inorganic semiconductors. A main difference between organic and inorganic disordered semiconductors is the shape of the density of states (DOS). In the inorganic semiconductors, the band gap states usually follow an exponential distribution. The energies of localized states in organic conductors are widely distributed due to several causes: the fluctuation of the lattice polarization energies, dipole interactions, and molecular geometry fluctuations (Bisquert & Quiñones, 2006).

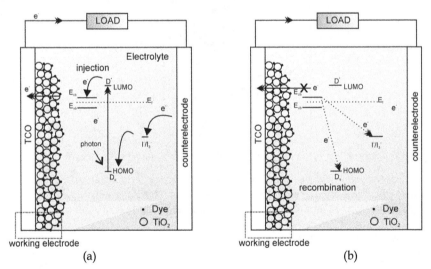

(a)   (b)

Fig. 2. (a) General schema for DSSCs, it is showed electron migration from dye molecule (D) through solar cell, (b) possible recombination process and, (c) Energy diagram for DSSCs.

Electron does a electric work at external load after go out from solar cell and then, come back through counterelectrode. Electrolyte transports the positive charges (holes) toward the counterelectrode and couple redox is reduced over its surface (eq. 3), on same time couple redox reduce the oxidized dye and regenerate the dye (eq. 4). Additionally In this process, some electrons can migrate from CB_{TiO2} to the HOMO level of the dye or electrolyte due to electron trapping effects; this process results in electron recombination (eq. 5, 6). These processes decrease the cell performance for affecting all its parameters (Fig. 2(b)); these process are presented because of differences on electron transfer rates between LUMO level CB_{TiO2} and the electron transfer rate into CB_{TiO2} (Kay & Grätzel 2002, S.S. Kim et. al 2003, Grätzel 2004).To achieve a cell efficient operation, the electron injection rate must be faster than the decay of the dye excited state. Also, the rate of reduction of the oxidized sensitizer (D⁺) by the electron donor in the electrolyte (eq. 4), must be higher than the rate of back reaction of the injected electrons with the dye cation (eq. 5), as well as the rate of reaction of injected electrons with the electron acceptor in the electrolyte (eq. 3). Finally, the kinetics of the reaction at the counter electrode must also guarantee the fast regeneration of redox couple (eq. 3) (Kalaignan & Kang, 2006). The oxidized dye must be regenerated by the redox couple at the speed of nanoseconds to kinetically compete with the metal oxide electrons for subsequent electron injection as well as to prevent the recombination, which depends on the energetic of metal oxide/dye/electrolyte interface (V. Thavasi et al, 2009).

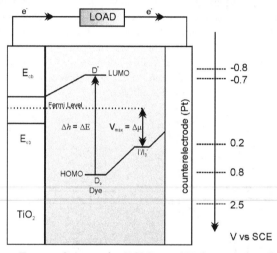

Fig. 3. General schema Energy diagram for DSSCs; and Voltage scale relative to SCE.

In terms of energetic levels shown in figure 3, electronic excitation in the dye (by light absorption) promotes the system into a high energy state, with associated electronic energy level, (LUMO level), simultaneously creating an electron deficient on low energy state (HOMO level). The electrons in these two states are separated by a difference in enthalpy (*h*), as follows (Bisquert et al, 2004):

$$\Delta h = \Delta E = E_{LUMO} - E_{HOMO} \tag{7}$$

$$\Delta E = E_C - E_V \text{ (in semiconductors)} \tag{8}$$

The departure of the population of the states from their thermal equilibrium values implies a difference in their chemical potential ($\Delta\mu$) as follows:

$$\Delta\mu = \mu_{LUMO} - \mu_{HOMO} \tag{9}$$

$$\Delta\mu = \mu_C - \mu_V \text{ (in semiconductors)} \tag{10}$$

Efficient operation of DSSCs relies on both efficient electron injection and efficient dye regeneration. Additionally, the LUMO energy level should be sufficiently higher than the

TiO_2 conduction band (E_{CB}) for efficient electron injection. And the $\Delta\mu$ of the redox couple should be higher than the HOMO energy level for efficient dye regeneration and sustained photocurrent production (see fig. 3). The maximum voltage of a DSSCs under illumination corresponds to the difference of the TiO_2 Fermi level (E_f) and the $\Delta\mu$ of the electrolyte (relative to Standard Calomel Electrode, see figure 2(c)). In the DSSCs, the voltage is between 0.6-0.8V, and currents between 16-25mA/cm^2 can be achieved under standard operating conditions; the world record on efficiency is 10.4% (Wang 2010, Green 2010).

3. Constituents of DSSCs

3.1 Working electrode (TiO₂)

TiO_2 thin films are extensively studied because of their interesting chemical, electrical and optical properties; TiO_2 film in anatase phase could accomplish the photocatalytic degradation of organic compounds under the radiation of UV. Therefore, it has a variety of application prospects in the field of environmental protection (Quiñones & Vallejo, 2010). TiO_2 thin film in rutile phase is known as a good blood compatibility material and can be used as artificial heart valves. In addition, TiO_2 films are important optical films due to their high reflective index and transparency over a wide spectral range (Mechiakh et. al, 2010). Despite the existence of other types of oxide semiconductor with high band-gap and band gap positions as SnO_2 and ZnO, the TiO_2 thin films are the most investigated material as photo-electrode to be used in DSSCs; because of the efficiency DSSCs constructed with TiO_2 electrodes yield the highest values of Isc, Voc, η and the IPCE (Bandaranayake, 2004). In this section we will review the main characteristics of TiO_2 thin film used as photoelectrode.

3.1.1 Structural properties of TiO₂

Titanium dioxide presents three mainly different crystalline structures: *rutile, anatase,* and *brookite* structures; and other structures as *cotunnite* has been synthesized at high pressures; in table 1 are listed some physical and chemical properties for three major structures of TiO_2. Despite their three stable structures, only rutile and anatase play any role on DSSCs, the unit cell for anatase and rutile structures are shown in figure (4a,b); in both structures the building block consists of a titanium atom surrounded by six oxygen atoms in a more or less distorted octahedral configuration. In each structure, the two bonds between the titanium and the oxygen atoms at the aspices of the octahedron are slightly longer (U. Diebold , 2003). In figure 5 is shown a typical X-ray diffraction pattern (XRD) of TiO_2 thin film (about 14μm of thickness) deposited by Atomic Pressure Chemical Vapor Deposition (APCVD) method; the bigger peak observed at 2θ=25.16° corresponds to preferential crystalline plane (110) of the crystalline phase anatase. It is important take into account that crystalline structure depends on deposition method and synthesis conditions used to obtain TiO_2 thin films; and different methods have been used to deposit TiO_2 thin films. Among these are the Sol–gel method by hydrolysis of $Ti(OiPr)_4$ followed by annealing at 500–600 °C, chemical vapor deposition (CVD), physical vapor deposition, chemical bath deposition (CBD), reactive sputtering and atomic layer deposition (ALD) (Quiñones & Vallejo, 2010).

3.1.2 Electrochemical properties of TiO₂

It is broadly accepted that DSSCs efficiency is mainly governed by the kinetics of charge transfer at the interface between TiO_2, the dye, and the hole transport material.

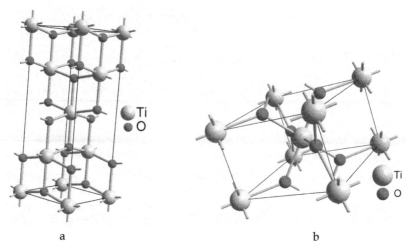

    a   b

Fig. 4. Unit cell for: (a) anatase TiO_2 and (b) rutile TiO_2.

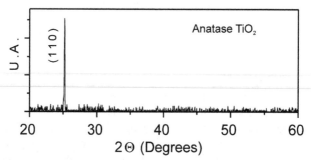

Fig. 5. XRD pattern of TiO_2 thin film, (take it on Shimadzu 6000 diffractometer with Cu-K$_\alpha$ radiation (λ = 0.15418 nm) source.

Crystal Structure	System	Space Group	Lattice constant (nm)			
			A	B	c	c/a
Rutile	Tetragonal	$D_{4h}^{14} - P4_2 / mnm$	0.4584	--	0.2953	0.644
Anatase	Tetragonal	$D_{4h}^{19} - I4_1 / amd$	0.3733	--	0.937	2.51
Brookite	Rhombohedral	$D_{2h}^{15} - Pbca$	0.5436	0.9166	0.5135	0.944

Density (Kg/m³)		Band gap Energy (eV)	Standard heat capacity (J/mol °C)
Rutile	4240	3.0 Indirect	55.06
Anatase	3830	3.2 Indirect	55.52
Brookite	4170	----	298.15

Table 1. TiO_2 bulk properties

The initial steps of charge separation in a DSSCs are the injection of an electron from a photoexcited dye to the conduction band of the TiO_2 and subsequently, the transfer of an electron from the hole transport molecule to the dye (Figure 2). The first process is usually completed within 200 ps, and the latter, the regeneration of the oxidized dye, is completed within the nanosecond time scale for liquid electrolyte DSSCs containing an (I^-/I_3^-) redox couple (Bisquert & Quiñones, 2006). It is very important to study this phenomenon with an appropriate analytical technique. Electrochemical impedance spectroscopy (EIS) is an experimental method of analyzing electrochemical systems; this method can be used to measure the internal impedances for the electrochemical system over a range of frequencies between mHz-MHz (Wang et al, 2005); additionally EIS allows obtaining equivalent circuits for the different electrochemical systems studied. Figure 6(a) shows a typical equivalent circuit for DSSCs; this model has four internal impedances. The first impedance signal (Z_1) related to the charge transfer at the platinum counter electrode in the high-frequency peak (in kHz range) and the sheet resistance (R_h) of the TCO in the high frequency range (over 1 MHz); the second signal (Z_2) related to the electron transport in the TiO_2/dye/electrolyte interface in the middle-frequency peak (in the 1–100 Hz), and the third signal (Z_3) related to the Nernstian diffusion within the electrolyte in the low-frequency peak (in the mHz range); in figure 6(b) is shown Nyquist diagram of a DSSCs from the result of a typical EIS analysis. Finally, the total internal impedance of the DSSCs is expressed as the sum of the resistance components (R_1, R_2, R_3, and R_h). High performance of the DSSCs is achieved when this total internal resistance is small (Shing et al, 2010).

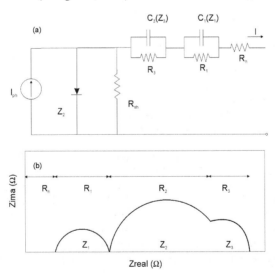

Fig. 6. Scheme of: (a) Equivalent circuit model for DSSCs and (b) Nyquist plot of the DSSCs from EIS analysis (adapted from Shin, 2010).

When you compared the equivalent circuit for DSSCs and conventional pn-junction solar cells, appears two large capacitance elements C_1 (in 10 µF/Cm^2 range) and C_3 (in 1F/Cm^2 range). Additionally R_{sh} can be described as (Islam & Han, 2006):

$$R_s = R_h + R_1 + R_3 \tag{11}$$

Under direct current conditions, the equivalent circuit for DSSCs is similar to conventional equivalent circuit for conventional solar cells, although the working mechanism is very different.

3.2 Sensitizer

One drawback with DSSCs is the high band gap of the core materials, e.g. TiO_2 particles <50–70nm in size (3.2 eV, wavelength <385nm in the ultraviolet range), compared to other semiconductors, which is illustrated by its absorption of UV light but not visible light. In the case of solar ultraviolet rays, only 2–3% of the sunlight (in the UV spectrum) can be utilized (Lee & Kang, 2010). To solve this, it appears the main characteristic of DSSCs: the dye, one of the most important constituents of the DSSCs. Some of the requirements that an efficient sensitizer has to fulfill include:

- A broad and strong absorption, preferably extending from the visible to the near-infrared.
- Minimal deactivation of its excited state through the emission of light or heat.
- Irreversible adsorption (chemisorption) to the surface of the semiconductor and a strong electronic coupling between its excited state and the semiconductor conduction band.
- Chemical stability in the ground as well as in the excited and oxidized states, so that the resulting DSSCs will be stable over many years of exposure to sun light.
- A reduction potential sufficiently higher (by 150–200mV) than the semiconductor conduction band edge in order to bring about an effective electroninjection.
- An oxidation potential sufficiently lower (by 200–300mV) than the redox potential of the electron mediator species, so that it can be regenerated rapidly.

A variety of transition-metal complexes and organic dyes has been successfully employed as sensitizers in DSSCs thus far, however the most efficient photosensitizers are ruthenium(II) polypyridyl complexes that yielded more than 11% sunlight to electrical power conversion efficiencies. The first two sensitizers used in DSSCs were N3 and N719. These are showed in figure 7. The molecular difference between these two compounds is the presence of tetrabutylammoniun ion to form two esters groups in N719. N3 and N719, affording DSSCs with overall power conversion efficiencies of 10.0% and 11.2 %respectively, harvest visible light very efficiently with their absorption threshold being at about 800 nm; figure 7 shows also the UV/VIS curve for N3 and N719 dyes, this profile is typical for al Rutenium complex used as dyes in DSSCs. Other important issue concerning dyes is their cost. Ruthenium, for instance, which is currently the most commonly utilized metal in metal-containing dyes for DSSCs, is a rare metal with a high price. Despite the fact that the contribution of the sensitizer to the total cell cost is limited, as efficient light harvesting requires a monolayer of sensitizer molecules, stable long-living ruthenium-based dyes are always highly desirable. Efficient, ruthenium-free sensitizers could also lead to such a cost decrease (Vougioukalakis et al, 2010).

Since the first Grätzel report, special attention has been paid to a number of details in the development of sensitizers in order to improve the photoelectric conversion efficiency and stability. A series of modifications of these early Ru(II) complexes have, among others, led to sensitizers with amphiphilic properties and/or extended conjugation. These amphiphilic new dyes try to achieve some of the following properties:

- A higher ground state pK_a of the binding moiety thus increasing electrostatic binding onto the TiO_2 surface at lower pH values.

- The decreased charge on the dye attenuating the electrostatic repulsion in between adsorbed dye units and thereby increasing the dye loading.
- Increasing the stability of solar cells towards water-induced dye desorption.
- The oxidation potential of these complexes is cathodically shifted compared to that of the N3 sensitizer, which increases the reversibility of the ruthenium III/II couple, leading to enhanced stability.

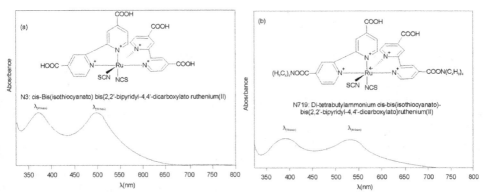

Fig. 7. Molecular structure of: (a) N3 and (b)N719.

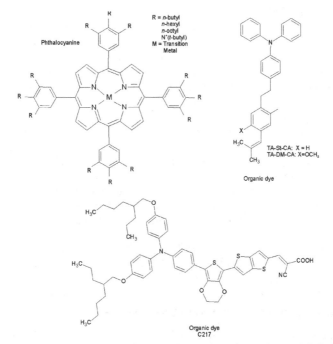

Fig. 8. The molecular structures of tetrakis(4-carboxyphenyl)porphyrin (Phthalocyanine) and two organic dyes (adapted from Vougioukalakis et al, 2010).

With respect to metal polypyridine complexes, organic dyes have been less investigated for sensitization of wide band gap semiconductors. Apart from the coumarin derivative, organic dyes such as porphyrins, phthalocyanines, perylene bis-amides, xanthenes and polyenes show low photon to electron conversion efficiencies. The understanding of the factors that determine these low performances could lead to the development of new efficient dyes, cheaper to manufacture than transition metal complexes (Argazzi et al, 2004). Figure 8 (a) shows some typical molecular structure of a basic phthalocyanine, (special polypyridine complexes); in this figure, R means a radical like: *t*-butl, *n*-hexil, *n*-octyl, in some cases R can be a substituted amine, and M means a metallic cation. Figure 8(b) shows some organic dye sensitizers used in DSSCs; TA-St-CA nad TA-TM-TA dyes contain a p-conjugated oligo-phenylenevinylene unit with an electron donor–acceptor moiety for intramolecular charge transfer, and a carboxyl group as an anchoring unit for the attachment of the dye onto TiO_2 nanoparticles; efficiences about 9% have been reported for DSSCs based in organic dyes. Additionally, C217 dye employing a lipophilic dihexyl oxy-substituted triphenylamine electrondonor, a cyanoacrylic electron acceptor, a distinguishable feature of this new amphiphilic D–π–A chromophore consists in a binary π-conjugated spacer. Here, an electron-rich 3,4-ethylenedioxythiophene unit is connected to electron-donor to lift the energy of the highest occupied molecular orbital (HOMO), while thienothiophene conjugated with A leads to a suitable lowest unoccupied molecular orbital (LUMO) energy; DSSCs based on C217 dye have reported efficiencies about 9.8%.

3.3 Electrolyte solution
3.3.1 Volatile Organic Solvents (VOS)

A typical DSSCs consists of a dye-coated mesoporous TiO_2 nano-particle film sandwiched between two conductive transparent electrodes, and a liquid electrolyte traditionally containing the (I^-/I_3^-) couple redox to fill the pores of the film and contact the nano-particle. The electrolyte, as one of the key components of the DCCS, provides internal electric ion conductivity by diffusing within the mesoporous TiO_2 layer and is an important factor for determining the cell performance (Wang, 2009). Volatile organic solvents (VOS) are commonly used as electrolyte (hole conductor), and nowadays, high efficiencies can be obtained for DSSCs using electrolytes based on VOS. When you compare the IPEC peak of a DSSCs construed with different types of electrolyte solutions, the biggest values are obtained with VOS; conventionally it IPEC peak value is about (74-78%) whilst others as ionic electrolytes present values about 53% and less solar efficiency. Typical VOS are acetonitrile, propionitrile, methoxyacetonitrile, and methoxypropionitrile, which present a high degree of ion conductivity, relatively high dielectric constant and ability to dissolve electrolytes; figure 9 shows the chemical strucuture of some typical VOS used in DSSCs.

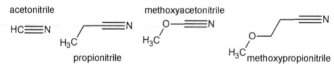

Fig. 9. Chemical strucuture of some typical VOS used in DSSCs.

Despite the relative high efficiency values of VOS in DSSCs (about 10%), the usage of these VOS electrolytes seriously limits a large scale implementation of this technology due to the poor long-term stability of the cells and the necessity of a complex sealing process (Wachter et al, 2010). Today this is one of the major drawbacks of DSSCs to implementation in a factory production line, and a new field research tries to replace the VOS in DSSCs by new type of electrolytes as: ionic liquid electrolytes, p-type inorganic semiconductors and polymer electrolytes. In principle, all materials with p-type semiconducting behavior, capable of accepting holes from the dye cation, are potential candidates to replace the liquid electrolyte in DSSCs. We will review main aspects of these possible substitutes.

3.3.2 Ionic liquid electrolytes

As an alternative to VOS, the use of room temperature ionic liquids (RTIL) and low molecular weight organic solvents with polar ligand have received considerable attention in recent decades. RTIL have several superior properties such as the non-volatility, chemical stability, not inflammability and high-ionic conductivity at room temperature. Nevertheless, the viscosity of the non-volatile solvents is usually higher than those of conventional volatile solvents, leading the energy conversion efficiency, which does not reach those of DSSCs employing conventional VOS electrolytes. The viscosity of typical ionic liquids is about 100 times larger than conventional VOS, and also 30 times larger than water at room temperature. Photocurrents in such systems are affected by the series resistances of the electrolytes, which are usually in proportion to the viscosity.

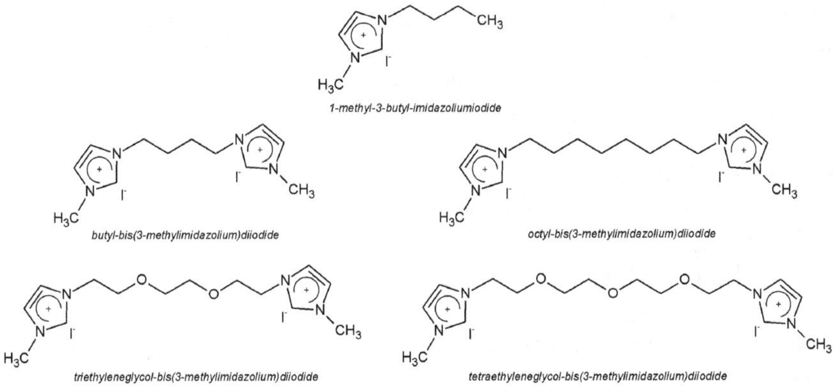

Fig. 10. Chemical strucuture of some typical ionic liquid electrolytes used in DSSCs (adapted from Zafer et al, 2009).

The low mass transport of redox couples is a limiting factor of applicability of these solvents in DSSCs. Therefore, high concentration of redox couples is required to obtain considerable conductivity in the electrolyte medium; however, in this case, the dark current may be increased owing to the high concentration of redox couples near to dye-attached to TiO_2 and the photocurrent generation is also interrupted by the absorption of visible light into the electrolyte medium (Kang et al, 2008). Currently, RTIL based on 1,3-dialkilimidazolium salts, have been studied as a electrolyte components for DSSCs. Relatively high viscosity of these type electrolytes appears to limit their applications. In order to improve charge transport properties, di-cationic bis-imidazolium based Ionic liquid electrolytes (ILs) with alkyl and polyether chains have been successfully investigated. Polyether groups improve the self organization of molecules in electrolyte which increase the charge transfer and the wetting capacity of the TiO_2 surface. Overall light-to-electric conversion efficiencies of 5.6%

have been achieved by dicationic bis-imidazolium iodide under simulated sunlight (Zafer et al 2009). It is of great interest that the fill factors of ionic liquid electrolytes-based DSSCs easily reach extremely high values of over 75% even in full sunlight. Dilution of the same ionic liquid with a low-viscosity organic solvent often decreases the fill factor. Figure 10 shows the chemical strucuture of some typical ionic liquids electrolytes used in DSSCs.

3.3.3 P-type semiconductors

The most common approach to fabricate solid-state DSSCs is by using p-type semiconductors. Several aspects are essential for any p-type semiconductor in a DSSCs:

- It must be able to transfer holes from the sensitizing dye after the dye has injected electrons into the TiO_2; that is, the upper edge of the valence band of p-type semiconductors must be located above the ground state level of the dye.
- It must be able to be deposited within the porous nanocrystalline layer.
- A method must be available for depositing the p-type semiconductors without dissolving or degrading the monolayer of dye on TiO_2 nanocrystallites.
- It must be transparent in the visible spectrum, or, if it absorbs light, it must be as efficient in electron injection as the dye.

Many inorganic p-type semiconductors satisfy several of the above requirements. However, the familiar large-band gap p-type semiconductors such as SiC and GaN are not suitable for use in DSSCs since the high-temperature deposition techniques for these materials will certainly degrade the dye. After extensive experimentation, a type of inorganic p-type semiconductor based on copper compounds such as CuI, CuBr, or CuSCN was found to meet all of these requirements (Li et al, 2006).

3.3.4 Polymer electrolytes

According to the physical state of the polymer, there are two types of polymer electrolytes used in DSSCs: (1) solid polymer electrolytes and (2) gel polymer electrolytes.

3.3.4.1 Solid polymer electrolytes

Compared with inorganic p-type semiconductors, organic p-type semiconductors (i.e. organic hole-transport materials) possess the advantages of having plentiful sources, easy film formation and low cost production. However, the conversion efficiencies of most of the solid-state DSSCs employing organic p-type semiconductors are relatively low particularly under high light irradiation. This is due to the low intrinsic conductivities of organic hole-transport material (HTMs), the high frequencies of charge recombination from TiO_2 to HTMs, and the poor electronic contact between dye molecules and HTMs, because of incomplete penetration of solid HTMs in the pores of the mesoporous TiO_2 electrodes. The ionic conductivity of solid polymer electrolytes is dependent on molar ratio of the polymer and the iodide salt due to the transfer efficiency of charge carriers and complex formation between metal cations and polymer atoms.

3.3.4.2 Gel polymer electrolytes (GPE)

Compared to other types of HTM, GPE are constructed by trapping liquid electrolytes, which usually contain organic solvents and inorganic salt ssuch as propylene (PC), Ehylene Carbonate (EC) or acetonitryle, lithium iodide (LiI), sodium iodide (NaI) and potassium iodide (KI). Compared to the corresponding parameters of the DSSCs based on liquid electrolytes, after gelation the decrease of J_{SC} (short-circuit density) due to the comparatively lower conductivity and the increase of VOC (open-circuit voltage) because of the suppression of dark current by polymer chains covering the surface of TiO_2 electrode result in the almost same efficiency for the DSSCs with GPE and with liquid electrolytes. Achieved by "trapping" a liquid electrolyte in polymer cages formed in a host matrix, GPE have some advantages, such as low vapor pressure, excellent contact in filling properties between the nanostructured electrode and counter-electrode, higher ionic conductivity compared to the conventional polymer electrolytes. Furthermore GPE possess excellent thermal stability and the DSSCs based on them exhibit outstanding stability to heat treatments. There was negligible loss in weight at temperatures of 200°C for ionic liquid-based electrolytes of poly (1-oligo(ethyleneglycol) methacrylate-3-methyl-imidazoliumchloride) (P(MOEMImCl). Thus the DSSCs based on GPE have outstanding long-term stability. Therefore, GPE have been attracting intensive attentions and these advantages lead to broad applications in the DSSCs Nowadays, several types of GPEs based on different types of polymers have already been used in the DSSCs, such as poly(acrolynitrile), poly(ethyleneglycol), poly(oligoethylene glycol methacrylate), poly(butylacrylate), the copolymers such as poly(siloxane-co-ethyleneoxide) and PVDF-HFP (Wang, 2009).

3.4 Redox couple

It is well known that the iodide salts play a key role in the ionic conductivity in DSSCs. Moreover, the basis for energy conversion is the injection of electrons from a photoexcited state of the dye sensitizer into the conduction band of the TiO_2 semiconductor on absorption of light. However, despite of its qualities; (I^-/I_3^-) couple redox has some drawbakcs, such as the corrosion of metallic grids (e.g., silver or vapor-deposited platinum) and the partial absorption of visible light near 430 nm by the I_3^- species. Another drawback of the (I^-/I_3^-) couple is the mismatch between the redox potentials in common DSSCs systems with Ru-based dyes, which results in an excessive driving force of 0.5~0.6 eV for the dye regeneration process. Because the energy loss incurred during dye regeneration is one of the main factors limiting the performance of DSSCs, the search for alternative redox mediators with a more positive redox potential than (I^-/I_3^-) couple is a current research topic of high priority. In order to minimize voltage losses, due to the Nernst potential of the iodine-based redox couple, and impede photocurrent leakage due to light absorption by triiodide ions, other redox couples have been also used, such as $SCN^-/(SCN)_3$; $SeCN^-/(SeCN)_3^-$, (Co^{2+}/Co^{3+}), (Co^+/Co^{2+}), coordination complexes, and organic mediators such as 2,2,6,6-tetramethyl-1-piperidyloxy (Min et al, 2010). Notwithstanding of different options and alternatives to replace (I^-/I_3^-) couple redox: this system presents highst solar cell efficiency. Additional, alternatives have been proposal to improve the efficiency of this type of DSSCs, as the adition of organic acid to electroylte solution or others aditives but until now best effiency has been reached with (I^-/I_3^-) couple redox.

3.5 Counter electrode

In DSSCs, counter-electrode is an important component, the open-circuit voltage is determined by the energetic difference between the Fermi-levels of the illuminated transparent conductor oxide (TCO) to the nano-crystalline TiO_2 film and the platinum counter-electrode where the couple redox is regenerated (McConnell, 2002). Platinum counter-electrode is usually TCO substrate coated with platinum thin film. The counter-electrode task is the reduction of the redox species used as a mediator in regenerating the sensitizer after electron injection, or collection of the holes from the hole conducting materials in DSSCs (Argazzi et al, 2004). Electrochemical impregnation from salts and physical deposition such as sputtering are commonly employed to deposit platinum thin films. Chemical reduction of readily available platinum salts such as H_2PtCl_6 or $Pt(NH_3)_4Cl_2$ by $NaBH_4$ is a common method used to obtain platinum electrodes. Platinum has been deposited over or into the polymer using the impregnation–reduction method (Yu et al, 2005). It is known that the final physical properties of Pt thin films depend on deposition method. Figure 11 shows SEM images of platinum thin films deposited by sputtering and electrochemical method as function of substrate type. Figure 11(a) corresponds to TCO substrate without platinum thin film, and Figure 11(b) shows a platinum thin film on TCO substrates deposited by electrochemical method. It is clear that TCO substrate grain size is smaller than platinum thin film grain size; this figure shows different size grain and Pt particles distribute randomly through out the substrate surface; this image shows some cracks in some places of the substrate. Furthermore, figure 11(c) shows platinum thin films deposited on TCO by sputtering method, it shows that platinum thin films have better uniformity than platinum thin film deposited by electrochemical method and the size grain is greater than size grain of thin film deposited by electrochemical method. In fig. 11(c) the Pt particles are distributed randomly through out the substrate without any crack; this is different to the electrochemical method, and indicates that the surface is uniformly coated. This thin film is less rough and corrects imperfections of substrate. Finally platinum thin film grown on glass SLG shows both smaller size grain particles and lower uniformity than platinum thin film deposited on TCO (Quiñones & Vallejo, 2011).

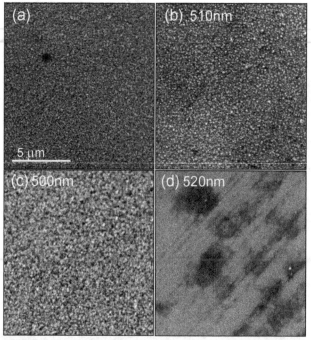

Fig. 11. SEM images (20000x) from: (a) TCO substrate; (b) Pt/TCO by electrochemical method; (c) Pt/TCO by sputtering; (d) Pt/SLG by sputtering (Quiñones & Vallejo et al, 2011).

Despite Pt has been usually used as counter electrode for the I_3^- reduction because of its high catalytic activity, high conductivity, and stability, Pt counter-electrode is one of the most expensive components in DSSCs. Therefore, development of inexpensive counter electrode materials to reduce production costs of DSSCs is much desirable. Several carbonaceous materials such as carbon nanotubes, activated carbon, graphite, carbon black and some metals have been successfully employed as catalysts for the counter electrodes. The results shows that carbonaceous materials not only gave ease in creating good physical contact with TiO_2 film but also functioned as efficient carrier collectors at the porous interface (Lei et al, 2010). Some possible substitutes to Pf thin films counter-electrode are:

3.5.1 Metal counter electrodes
Metal substrates such as steel and nickel are difficult to employ for liquid type DSSCs because the $I-/I_3-$ redox species in the electrolyte are corrosive for these metals. However, if these surfaces are covered completely with anti-corrosion materials such as carbon or fluorine-doped SnO_2, it is possible to employ these materials as counter-electrodes. Metal could be beneficial to obtain a high fill factor for large scale DSSCs due to their low sheet resistance. Efficiencies around 5.2% have been reported for DSSCs using a Pt-covered stainless steel and nickel as counter-electrode (Murakami & Grätzel, 2008).

3.5.2 Carbon counter electrode
First report of carbon material as counter electrode in DSSCs was done by Kay and Grätzel. In this report they achieved conversion efficiency about 6.7% using a monolithic DSSCs embodiment based on a mixture of graphite and carbon black as counter electrode (Kay & Grätzel, 1997).

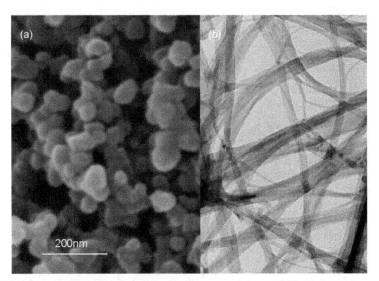

Fig. 12. SEM images (30000x) from: (a) Carbon nanoparticles and (b) Carbon nanotubes.

The graphite increases the lateral conductivity of the counter electrode and it is known that carbon acts like a catalyst for the reaction of couple redox ($I_3-/I-$) occurring at the counter electrode. Recently, carbon nanotubes have been introduced as one new material for counter electrodes to improve the performance of DSSCs (Gagliardi et al, 2009). The possibility to obtain nanoparticles and nanotubes of carbon permits investigation of different configuration in synthesis of counter electrode fabrication, to improve the DSSCs efficiency. In figure 12 are showing the scanning electron microsocopy images of nanoparticles and nanotubes of carbon.

4. Efficiency and prospects

From first Grätzel report, the efficiency of DSSCs with nano-porous TiO_2 has not changed significantly. Currently, the world record efficiency conversion for DSSCs is around 10.4% to a solar cell of 1 cm² of area; in table 2 are shown the confirmed efficiencies for DSSCs.

The high efficiency (table 2) of DSSCs has promoted that many institutes and companies developed a commercial research on up-scaling technology of this technology. The Gifu University in Japan, developed colorful cells based on indoline dye and deposited with zinc oxide on large size of plastic substrate. The Toin University of Yokohama in Japan fabricated the full-plastic DSSCs modules based on low-temperature coating techniques of TiO_2 photoelectrode. Peccell Technologies in Japan, and Konarka in US, practiced the utility and commercialization study about flexible DSSCs module on polymer substrate. Léclanche S.A, in Switzerland, developed outer-door production of DSSCs. INAP in Germany gained an efficiency of 6.8% on a 400 cm² DSSCs module. However, despite prospective of DSSCs technology, the degradation and stability of the DSSCs are crucial topics to DSSCs up-scaling to an industrial production (Wang et al, 2010).

Device	Efficiency (%)	Area (cm²)	V_{oc} (V)	J_{sc} (mA)	FF	Test center
DSSCs$_{cell}$	11.2 +/-0.3	0.219	0.738	21	72.2	AIST*
DSSCs$_{cell}$	10.4+/-0.3	1.004	0.729	22.0	65.2	AIST
DSSCs$_{submodule}$	9.9+/-0.4	17.11	0.719	19.4	74.1	AIST

*Japanese National Institute of Advanced Industrial Science and technology.

Table 2. Confirmed terrestrial DSSCs efficiencies measured under the global AM1.5 spectrum (1000W/m²) at 25°C (Green et al 2010).

According to the operation principle, preparation technology and materials characteristics of DSSCs, they are susceptible to:

- Physical degradation: the system contains organic liquids which can leak out the cells or evaporate at elevated temperatures. This could be overcome using appropriate sealing materials and low volatiles solvents.
- Chemical degradation: The dye and electrolyte will photochemically react or thermal degrade under working conditions of high temperature, high humidity, and illumination. The performance DSSCs will irreversiblely decrease during the process causing the life time lower than commercial requirements (>20 years)

Unfortunately, there are not international standards specific in DSSCs. Nowadays, most of the performance evaluation of DSSCs is done according to International electrotechnical commission (IEC), norms (IEC-61646 and IEC-61215), prepared for testing of thin film photovoltaic modules and crystalline silicon solar cells. Most of the on up-scaling technology was made with these IEC international standards (Wang et al, 2010).

5. Conclusion

In this Chapter, the physics and chemistry of the dye sensitizer solar cells were reviewed using own studies and some of the last reports in the area. Different aspects related with basic principle and developments of each component of the solar cell was presented. This type of technology presents different advantages with its homologues inorganic solar cells, and nowadays DSSCs are considered one economical and technological competitor to pn-junction solar cells. This technology offers the prospective of very low cost fabrication and easy industry introduction. However, the module efficiency of DSSCs needs to be improved to be used in practical applications. It is necessary to achieve the optimization of the production process to fabricate photoelectrodes with high surface area and low structural defects. It is necessary to solve problems asociated to encapsulation of (I^-/I_3^-) redox couple in an appropiate medium such Ionic liquid electrolytes, p-type semiconductors, Solid polymer electrolytes, Gel polymer electrolytes and the deposited stable and cheap counterelectrode. If this problems are solved is possible that in near future DSSCs technology will become in a common electrical energy source and widely used around the world.

6. References

O'Regan, M. Grätzel (1991). A low cost high efficiency solar cell based on dye sensitized colloidal TiO_2. *Nature*, Vol. 353, pp. 737–740.

Donald Neamen (1997). *Semiconductor Physics and Deivices*, Second Edition, Mc Graw Hill, pp. 130.

S. Pooman, R. M. Mehra (2007). Effect of electrolytes on the photovoltaic performance of a hybrid dye ZnO solar cell. *Solar Energy Materials & Solar Cells*, Vol. 91, pp. 518-524.

C.A. Kelly, G.J. Meyer (2001). Excited state processes at sensitized nanocrystalline thin film semiconductor interfaces. *Coordination Chemistry Reviews*. Vol. 211, pp. 295.

M. Grätzel (2004). Conversion of sunlight to electric power by nanocrystalline dye-sensitized solar cells. *Journal of Photochemistry and Photobiology A: Chemistry*. Vol. 164, pp. 3–14.

A. Kay, M. Grätzel (2002). Dye-Sensitized Core-Shell Nanocrystals: Improved Efficiency of mesoporous Tin Oxide Electrodes Coated with a thin layer of an Isulating Oxide. *Chemistry of Materials*, Vol. 14 pp. 2930.

S.S. Kim, J.H. Yum, Y.E. Sung (2003). Improved performance of a dye-sensitized solar cell using a TiO_2/ZnO/Eosin Y electrode. *Solar Energy Materials & Solar Cells*. Vol. 79, pp. 495.

G. Kalaignan, Y. Kang (2006). *Journal of Photochemistry and Photobiology C: Photochemistry Reviews*, Vol. 7 pp. 17–22.

V. Thavasi, V. Renugopalakrishnan, R. Jose, S. Ramakrishna (2009). Controlled electron injection and transport at materials interfaces in dye sensitized solar cells. *Materials Science and Engineering: Reports*, Vol. 63 pp. 81–99.

J. Bisquert, D. Cahen, G. Hodes, S. Rühle, A. Zaban, (2004). Physical Chemical Principles of Photovoltaic Conversion with Nanoparticulate, Mesoporous Dye-Sensitized Solar Cells. *Journal of Physical Chemistry B*, Vol. 108 pp. 8106

J. Bisquert, E. Palomares, C. Quiñones, (2006). Effect of Energy Disorder in Interfacial Kinetics of Dye-Sensitized Solar Cells with Organic Hole Transport Material. *The Journal of Physical Chemistry B*, Vol. 110, *pp.* 19406-19411.

M. Wang, J. Liu, N. Cevey-Ha, S. Moon, P. Liska, R. Humphry-Baker, J. Moser, C. Grätzel, P. Wang, S. Zakeeruddin, M Grätzel (2010). High Efficiency solid-state sensitized heterojunction photovoltaic device. *Nano Today*, Vol. 5, pp. 169-174.

M. A. Green, K. Emery, Y. Hishikawa, W. Warta (2011). Solar cell efficiency tables (version 37). *Progress in Photovoltaics: Research and Applications*. Vol. 19, pp. 84.

C. Quiñones, J. Ayala, W. Vallejo (2010). Methylene blue photoelectrodegradation under UV irradiation on Au/Pd-modified TiO_2 films. Applied Surface Science, Vol. 257, pp. 367-371

C. Quiñones, W. Vallejo, G. Gordillo (2010). Structural, optical and electrochemical properties of TiO2 thin films grown by APCVD method'. *Applied Surface Science* Vol. 256, pp. 4065-4071.

R. Mechiakh, N. Ben Sedrine, R. Chtourou, R. Bensaha (2010). Correlation between microstructure and optical properties of nano-crystalline TiO_2 thin films prepared by sol–gel dip coating. *Applied Surface Science*, Vol. 257 pp. 670–676.

K.M.P. Bandaranayake, M.K. I. Senevirathna, P. Weligamuwa, K. Tennakone (2004). Dye-sensitized solar cells made from nanocrystalline TiO_2 films coated with outer layers of different oxide materials. *Coordination Chemistry Reviews*, Vol 248, pp. 1277-1281.

U. Diebold (2003). The Surface Science of TiO_2. *Surface Science Reports*, Vol. 48, pp. 53.

Bin Li, Liduo Wang, Bonan Kang, Peng Wang, Yong Qiu (2006). 'Review of recent progress in solid-state dye-sensitized solar cells'. *Solar Energy Materials & Solar Cells*, Vol. 90 pp. 549-573.

Y. Lee, M. Kang (2010). Comparison of the photovoltaic efficiency on DSSCs for nanometer sized TiO2 using a conventional sol–gel and solvothermal methods. *Journal of Industrial and Engineering Chemistry*, Vol. 122, pp. 284-289.

P. Wachter, M. Zistler, C. Schreiner, M. Berginc, U. O. Krasovec, D. Gerhard, P. Wasserscheid, A. Hinsch, H. J. Gores (2008). *Journal of Photochemistry and Photobiology A: Chemistry*, Vol. 197, pp. 25-33.

B. Lei, W. Fang, Y. Hou, J. Liao, D. Kuang, C. Su (2010). Characterisation of DSSCs-electrolytes based on 1-ethyl-3-methylimidazolium dicyanamide: Measurement of triiodide diffusion coefficient, viscosity, and photovoltaic performance'. *Journal of Photochemistry and Photobiology A: Chemistry*, Vol. 216, pp. 8-14.

Y. Wang (2009). Recent research progress on polymer electrolytes for dye-sensitized solar cells. *Solar Energy Materials & Solar Cells*, Vol. 93, pp. 1167-1175.

A. Luque, S. Hegedus. *Handbook of Photovoltaic Science and Enginnering*. John Wiley & Sons. USA. 2005.

M. Kang, K. Ahn, J. Lee, Y. Kang. Dye-sensitized solar cells employing non-volatile electrolytes based on oligomer solvent. *Journal of Photochemistry and Photobiology A: Chemistry* 195 (2008) 198-204

C. Zafer, K. Ocakoglu, C. Ozsoy, S. Icli. Dicationic bis-imidazolium molten salts for efficient dye sensitized solar cells: Synthesis and photovoltaic properties. *Electrochimica Acta* 54 (2009) 5709-5714.

A. J. Frank, N. Kopidakis, J. V. Lagemaat (2004). Electrons in nanostructured TiO_2 solar cells: transport, recombination and photovoltaic properties. *Coordination Chemistry Reviews*, Vol. 248, pp. 1165-1179.

J. Min, J. Won, Y.S. Kang, S. Nagase (2010). Benzimidazole derivatives in the electrolyte of new-generation organic dyesensitized solar cells with an iodine-free redox mediator. *Journal of Photochemistry and Photobiology A: Chemistry* (2010). doi:10.1016/j.jphotochem.2011.02.004

R. Argazzi, N. Iha, H. Zabri,F. Odobel, C. Bignozzi (2004). Contributions to the development of ruthenium-based sensitizers for dye-sensitized solar cells. *Coordination Chemistry Reviews*, Vol. 248, pp. 1299-1316.

P. Yu, J. Yan, J. Zhang, L. Mao. Cost-effective electrodeposition of platinum nanoparticles with ionic liquid droplet confined onto electrode surface as micro-media. *Electrochemistry Communications*, Vol. 9, pp. 1139-1144.

R.D. McConnell (2002). Assessment of the dyesensitized solar cell. *Renewable & Sustainable Energy Reviews*, Vol. 6, pp. 273-295.

G. C. Vougioukalakis, A.Philippopoulos, T. Stergiopoulos, P. Falaras (2010). Contributions to the development of ruthenium-based sensitizers for dye-sensitized solar cells. *Coordination Chemistry Reviews* (2010),doi:10.1016/j.ccr.2010.11.006

T. N. Murakami, M. Grätzel (2008). Counter electrodes for DSC: Application of functional materials as catalysts. *Inorganica Chimica Acta*, Vol. 361 pp. 572-580

A. Kay, M. Grätzel. On the relevance of mass transport in thin layer nanocrystalline photoelectrochemical solar cells. *Solar Energy Materials & Solar Cells*, Vol. 44, pp. 99.

S. Gagliardi, L. Giorgi, R. Giorgi, N. Lisi, Th. D. Makris, E. Salernitano, A. Rufoloni (2009). Impedance analysis of nanocarbon DSSCs electrodes. *Superlattices and Microstructures*, Vol. 46, pp. 205-208.

R. Argazzi, N. Y. M. Iha, H. Zabri, F. Odobel, C. A. Bignozzi (2004). Design of molecular dyes for application in photoelectrochemical and electrochromic devices based on nanocrystalline metal oxide semiconductors. *Coordination Chemistry Reviews*, Vol. 248, pp. 1299-1316.

Q.Wang, J. Moser, M. Grätzel (2005). Electrochemical Impedance Spectroscopic Analysis of Dye-Sensitized Solar Cells. *Journal of Physical Chemistry B*, Vol. 109, pp. 14945-14953

I. Shin, H. Seo, M. Son, J. Kim, K. Prabakar, H. Kim (2010). Analysis of TiO_2 thickness effect on characteristic of a dye-sensitized solar cell by using electrochemical impedance spectroscopy'. *Current Applied Physics* Vol. 10, pp. 422-424

C. Zafer, K. Ocakoglu, C. Ozsoy, S. Icli (2009). Dicationic bis-imidazolium molten salts for efficient dye sensitized solar cells: Synthesis and photovoltaic properties. *Electrochemica Acta*, Vol. 54, pp. 5709-5714.

N. K. A. Islam, Y. C. L. Han (2006). Improvement of efficiency of dye-sensitized solar cells based on analysis of equivalent circuit. *Journal of Photochemical and Photobiology A Chemistry*, Vol. 182, pp. 296-305.

L. Wang, X. Fang, Z. Zhang (2010). Design methods for large scale dye-sensitized solar modules and the progress of stability research'. Renewable and Sustainable Energy Reviews, Vol. 14, pp. 3178-3184.

C. Quiñones, W. Vallejo, F. Mesa (2010). Physical and electrochemical study of platinum thin films deposited by sputtering and electrochemical methods, *Applied Surface Science* 257 (2011) 7545–7550

Preparation of Hollow Titanium Dioxide Shell Thin Films from Aqueous Solution of Ti-Lactate Complex for Dye-Sensitized Solar Cells

Masaya Chigane, Mitsuru Watanabe and Tsutomu Shinagawa
Osaka Municipal Technical Research Institute
Japan

1. Introduction

As photovoltaic devices possessing potential for low processing costs and flexible architectures, dye-sensitized solar cells (DSSCs) using nanocrystalline TiO_2 (nc-TiO_2) electrodes have been extensively studied.(Bisquert et al., 2004; O'Regan & Grätzel, 1991) Congruently with increasingly urgent dissemination of solar cells against crisis of a depletion of fossil fuel, DSSCs are as promising alternative to conventional silicon-type solar cells. The main trend of investigations of DSSCs originates from the epoch-making works by Grätzel and co-workers in the early 1990s. (O'Regan & Grätzel, 1991) A typical construction of the cells are composed of dye-molecules (usually Ru complexes) coated nc-TiO_2 electrodes on transparent-conductive (TC) backcontact (usually fluorine-doped tin oxide (FTO)) glass substrate and counter Pt electrodes sandwitching triiodine/iodine [I_3^-/I^-] redox liquid electrolyte layer maintaining electrical connection with the counter Pt electrode. The voids of the network of TiO_2 nanoparticles connection form nanopores which are efficiently filled with electrolyte solution. An operation mechanism of DSSC begins with harvesting incident light by dye-molecules via photoexcitation of electron from the highest occupied molecular orbital (HOMO) to the lowest unoccupied molecular orbital (LUMO). The photoexcited electrons are transferred to the conduction band of the nc-TiO_2 and diffuse in TiO_2 matrix to TC layer followed by ejection to outer electric load. The oxidized dye is reduced by the electrolyte (I^-) and the positive charge is transported to Pt counter electrode. As well as close fitting of photo-absorption spectra of dyes to the spectrum of sunlight mainly in visible light region (nearly panchromatic dyes) (Nazeeruddin et al., 2001) the strong dye-TiO_2 coupling leading to rapid electron transfer from excited dye to TiO_2 (Tachibana et al., 1996) realizes practically promising solar-to-electrical conversion efficiency: more than 10 %. The charge separation of DSSCs occurs at the interface TiO_2 nanoparticles / dye molecules / [I^-/I_3^-] electrolyte. Therefore the combination of Ru-complex and TiO_2 is currently almost ideal choice in DSSC. Some problems of the TiO_2 nanoparticles electrode, however, remain room to investigate. Connection points of TiO_2 nanoparticles decrease an effective area of interface, and play a role on electron scattering sites, leading to restrict the conversion efficiency.(Enright & Fizmaurice, 1996; Peng et al., 2003) Though denser films seemingly improve the electron migration, they result in decrease of surface area for dye adsorption. Additionally TiO_2 nanoparticles electrodes are usually prepared by embrocation methods, e. g., a squeegee method, whereas via these methods great amount of Ti resource is consumed.

For the settlement several nanostructures of TiO$_2$ electrodes for DSSCs containing the array of nanorods,(Kang et al., 2008) nanotubes (Kang et al., 2009; Paulose et al., 2008) and assembly of spherical hollow (Kondo et al., 2006) or hemispherical (Yang et al., 2008) shells particles have been proposed owing to their ordered structures leading to ordered electron transport and large surface area for small amount of titanium as depicted in Fig. 1.

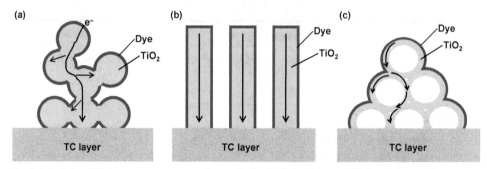

Fig. 1. Models of TiO$_2$ nano-structure electrode for DSSCs, (a) standard nanocrystalline particles, (b) nano tube or nano pillar arrays and (c) hollow shells.

Some works on ordered and multilayered hollow TiO$_2$ shells, which are inverse opal structure, have shown photonic crystalline effects leading to red shift in incident photon-to-current conversion efficiency (IPCE). (Nishimura et al., 2003; Yip et al., 2008) Recently energy conversion efficiencies of DSSCs using inverse TiO$_2$ opal, including 1.8 %, (Guldin et al., 2010) 3.47 % (Kwak et al., 2009) and 4.5 % (Qi et al., 2009) have been reported. In the previous work we prepared hollow TiO$_2$ shell monolayer films by the electrolysis of an aqueous (NH$_4$)$_2$TiF$_6$ solution on complicated polystyrene (PS) particles-preadsorbed substrate followed by calcination. (Chigane et al., 2009) Among few papers (Karuppuchamy et al., 2001; Yamaguchi et al., 2005) reporting the electrolytic preparation of TiO$_2$ for DSSC anode, the previous work first reported the DSSC conversion efficiency (0.63 %) using TiO$_2$ film prepared via electrolysis to our knowledge. From standpoint of methodology for a preparation of TiO$_2$ films electrolyses (electrodeposition) from aqueous solutions are a low cost and low resource consuming fabrication techniques since the deposition reaction occurs only nearby substrate. The (NH$_4$)$_2$TiF$_6$ solution is stable for long term, being able to undergo repeated electrolyses. However some industrial problems: liberation of highly toxic F- during electrochemical deposition reaction leading to bad working environment. Moreover insufficient conversion efficiency calls multilayered hollow structures. As a water-soluble and environment-benign titanium compound titanium bis(ammonium lactato)dihydroxide (TALH) increasingly attracts attention. (Caruso et al., 2001; Rouse & Ferguson, 2002) Especially Ruani and co-workers (Ruani et al., 2008) have developed single-step preparation of PS-TALH core-shell precursor from a suspension containing both PS and TALH, followed by fabrication inverse opal TiO$_2$ films by calcination. The process seems to be simple and time-saving compared with other conventional methods: PS template followed by infiltration of Ti oxides or Ti compounds sol. (Galusha et al., 2008; King et al., 2005; Liu et al., 2010; Nishimura et al., 2002) However DSSC electrode properties of the film have not been

Preparation of Hollow Titanium Dioxide Shell Thin Films from Aqueous Solution of Ti-Lactate Complex for
Dye-Sensitized Solar Cells

155

evaluated. There assumably are two reasons that the films becomes noncontiguous by calcination owing to drastic volume change from Ti-lactate to oxide and that H-TiO$_2$ film are difficult to be prepared in wide area. The latter problem is mainly due to aninonic surface functional groups of PS in usual cases despite electrostatically repulsive Ti-lactate anion leading to difficult formation of homogeneous structure. Based upon these trend and problems we propose in the present article some preparation methods of three-dimensional assembly of H-TiO$_2$ shells being applied to DSSC. Figure 2 illustrates our preparation process of H-TiO$_2$ shell films.

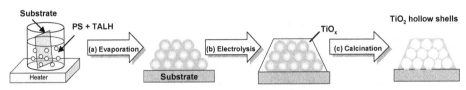

Fig. 2. Schematic representation of preparation process of hollow TiO$_2$ shell film.

The immersion of substrate in the initial suspension which contains both PS and TALH, followed by evaporation of water, forms a PS template coated with TALH (PS-TALH) precursor (Fig. 2 (a)). So as to avoid the volume change of films we supported the PS-TALH by electrodeposition of titanium oxide (TiO$_x$) thin film (Fig. 2(b)) thereon. The novelty of this method includes i) employing non-toxic TALH and PS with cationic surface groups which are supposedly good affinity with each other and ii) electrolysis of TALH solution for TiO$_x$ coverage. As a whole, this article highlights a low cost, facile and soft fabrication process of hollow TiO$_2$ (H-TiO$_2$) shell films and aggressive participation of them in DSSC as a trendy nano-architechtural electrode.

2. Experimentals

Two types of PS possessing anionic (A-PS) and cationic (C-PS) functional group on the surface, prepared by an emulsifier-free emulsion polymerization of styrene monomer with potassium persulfate (KPS) and 2,2'-azobis(2-methylpropionamidine) dihydrochloride (AIBA) as a radical initiator, respectively, were used.(Watanabe et al., 2007) From SEM observations a diameter of A-PS and C-PS beads was ca. 400 nm and ca. 300 nm, respectively. A TC glass plate coated with fluorine-doped tin oxide (FTO, 10 Ω/square, ASAHI GLASS Co., Ltd, A11DU80) and quartz glass plate (1 mm of thickness) were used as substrates. As pretreatments of substrates, FTO (15 mm × 20 mm) was degreased by anodic polarization (+5 mA cm^{-2}) against Pt counter electrode for 30 s in a 1 mol dm^{-3} NaOH and quartz glass plate (20 mm × 40 mm) was immersed in a 10 % NaOH solution for 10 min at 333 K. Both substrates were thoroughly rinsed with deionized water and immersed in the colloidal suspension of PS and TALH in a cylindrical glass bottle being bent backward at about 60° against bottom. (Hartsuiker & Vos, 2008; Ye et al., 2001) The glass bottle containing the suspension and substrate was placed on a hot plate the temperature of surface of which was set at 345 K. In this way the temperature of the suspension was maintained at 325 K for 5-24 h until complete evaporation of water. The initial concentrations of PS and TALH in the suspension were 0.28 % and 0.0025 or 0.005 mol dm^{-3}, respectively. Hollow shells TiO$_2$ (H-TiO$_2$) films were formed by the calcinations of the PS-TALH precursor at 723 K for 1 h. In the calcination, the temperature was raised in rate of 2 h from room temperature to 723 K. In some cases TiO_x films were electrodposited on the PS-TALH precursor films before calcination by a cathodic electrolysis in the electrolyte solution containing a 0.05 mol dm^{-3} TALH (Aldrich; reagent grade) and a 0.1 mol dm^{-3} NH_4NO_3 at -2 mA cm^{-2} of current density for 6 C cm^{-2} of charge density.

X-ray diffraction (XRD) patterns of the films in 3 cm^2 of deposition area were recorded on a RIGAKU RINT 2500 diffractometer (Cu $K\alpha$; λ = 0.1541 nm; 40 kV; 100 mA), with the incident angle (θ) fixed at 1°, scanning the diffraction angle (2θ) stepwise by 0.05° with a counting time of 10 s.

Transmission spectra and relative diffuse reflection (DR) spectra in ultraviolet (UV)–visible range of hollow shells were measured by means of Shimadzu UV-3150 spectrometer. An integral spherical detector equipment (Shimadzu ISR-3100) was used for DR spectroscopy with $BaSO_4$ powder (Wako Pure Chemical Industries) as a reference reflector. For the powder sample the hollow shell film was detached from quartz glass substrate by scratching with spatula.

X-ray photoelectron (XP) spectra of the films were obtained by means of Kratos AXIS-ULTRA DLD. A monochromated Al $K\alpha$ (1486.6 eV; 150 W) line was used as the X-ray source. The pressure in the analyzing chamber was lower than $< 1 \times 10^{-8}$ Torr during measurements. Binding energies of Ti 2p and O 1s photoelectron peaks were corrected from the charge effect by referencing the C 1s signal of adventitious contamination hydrocarbon to be 284.8 eV.

For DSSC measurements, a composite TiO_2 (C-TiO_2) film composed of flat bottom TiO_2 layer and hollow shell film was prepared. The deposition area, corresponding to an active area of the cell, was adjusted to be 0.25 cm^2 using a mask. Initially TiO_x film was galvanostatically electrodeposited on the FTO in the electrolyte solution containing a 0.05 mol dm^{-3} TALH (Aldrich; reagent grade) and a 0.1 mol dm^{-3} NH_4NO_3 at -3 mA cm^{-2} of current density for 10 C cm^{-2} of charge density as a blocking layer of DSSC electrode. On the FTO/TiO_x layer the PS-TALH precursor was coated from aqueous suspension of 0.28 % C-PS and 0.0025 mol dm^{-3} TALH and thereon TiOx was coated by electrolysis. The TiO_2 hollow films on FTO substrates were immersed in an ethanolic solution of 0.3 mmol dm^{-3} ruthenium dye (bistetrabutylammonium cis-di(thiocyanato)-bis(2,2'-bipyridine-4-carboxylic acid, 4'-carboxylate ruthenium(II), Solaronix N719) for 14–16 h at room temperature. An electrolyte solution for DSSC was composed of 0.1 mol dm^{-3} LiI, 0.05 mol dm^{-3} I_2, 0.6 mol dm^{-3} 1,2-dimethyl-3-propylimidazolium iodine (DMPII, Solaronix) and 0.5 mol dm^{-3} 4-tert-butylpyridine in acetonitrile. A platinum-coated glass substrate was used for a counter electrode. The dye-coated TiO_2 electrode and the counter electrode were set sandwiching a separation polymer sheet (25 µm of thickness; Solaronix SX-1170) and the electrolyte. Photovoltaic current density (J)-voltage (V) curves were obtained using an instrument for the measurements of solar cell parameters (Bunkoh-Keiki Co., Ltd K0208, with a Keithley 2400) with photoirradiation by a 150 W xenon lamp under the condition that was simulated airmass 1.5 solar irradiance with the intensity of 100 mW cm^{-2}. Incident photon-to-electron conversion efficiency (IPCE) spectra ranging 400 to 800 nm of wavelength were measured by means of a spectral photosensitivity measurement system (Bunkoh-Keiki) with a 150 W xenon lamp light source. Calibration was performed using a standard silicon photodiode (Hamamatsu Photonics, S1337-1010BQ).

As a reference sample, TiO_2 nanocrystalline particles (nc-TiO_2) electrode prepared by a squeegee method from TiO_2 colloidal solution (Solaronix Ti-Nanoxide D) and calcination in the same way as the hollow films was subjected to DSSC measurements.

The area density: amount of deposited TiO_2 against area (mg cm^{-2}) was determined using an
X-ray fluorescent spectrometer (Rigaku RIX3100).

3. Results and discussion

3.1 Characterization of H-TiO$_2$ films

Figure 3 shows SEM photographs of PS-TALH precursors. For both A-PS and C-PS, the sizes
of spherical PS-TALH units appeared to be almost same as PS itself and core/shell
structures of PS/TALH were not clearly observed as shown in expanded images (Fig. 3(b)
and (d)). Some bulgy junctions between adjacent spheres, however, suggest accumulation of
TALH. From appearance with eyes and SEM images, PS particles and TALH were
uniformly mixed.

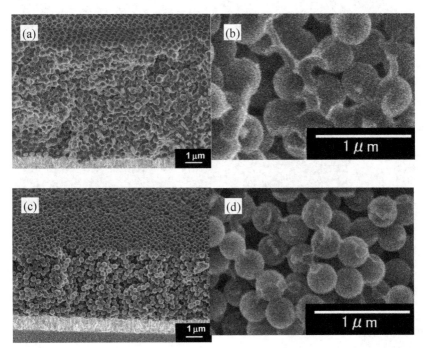

Fig. 3. Cross-sectional SEM photographs of PS-TALH films on FTO substrates from (a) or (b)
A-PS and (c) or (d) C-PS, respectively.

Figure 4 shows SEM photographs of H-TiO$_2$ films. More than 20 layers of quasi-ordered
spherical hollow shells were observed at some view points. The thickness of shell wall
was *ca.* 10 nm. For the sample started from suspension containing A-PS and TALH 0.0025
mol dm^{-3}, the fcc configuration was observed especially on the surface in one to several
millimeters range. However, the quasi-fcc ordered shells were discretely scattered on the
substrate, and some parts of the substrate were not covered with films. On the other hand,
although for the sample from C-PS the ordered structure was not observed (Fig. 4(d)) in
micrometer view, the hollow shells smoothly coated almost all over the substrate. This
more excellent dispersion by C-PS in appearance is due to uniform mixture of PS and

TALH in the precursor originating from their affinitive relationship. In both cases, there are observed many broken points of shells owing to PS-PS spheres contact at the initial precursor as shown in an inserted scheme in Fig. 4, leading to skeleton-like structure after calcinations.

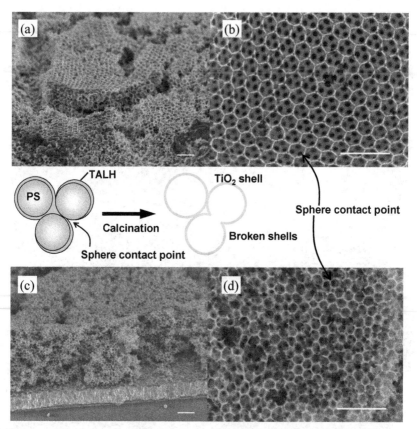

Fig. 4. Cross-sectional or suface SEM photographs of H-TiO$_2$ films on FTO substrates prepared from (a) or (b) A-PS and (c) or (d) C-PS, respectively. Scale bars correspond 1 μm.

Figure 5 shows XRD pattern of hollow shells film prepared on a quartz glass substrate by calcination of C-PS-TALH precursor. The pattern shows the TiO$_2$ shell film to be predominantly anatase type crystalline phase while the peaks were broad owing to nanocrystalline.(International Center for Diffraction Data, 1990)

The crystallite size (D) of the calcined film can be estimated from Scherrer's formula (Kim et al., 2008)

$$D = \frac{0.89\lambda}{\beta\cos\theta} \qquad (1)$$

where λ is the wavelength of the X-ray, β is the peak width and θ is the Bragg angle of the peak.

Preparation of Hollow Titanium Dioxide Shell Thin Films from Aqueous Solution of Ti-Lactate Complex for
Dye-Sensitized Solar Cells

159

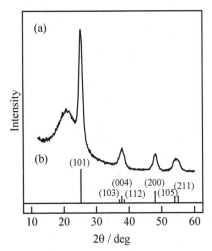

Fig. 5. XRD patterns of (a) H-TiO$_2$ film on quartz glass substrate and (b) an authentic pattern
of anatase-type TiO$_2$.(International Center for Diffraction Data, 1990)

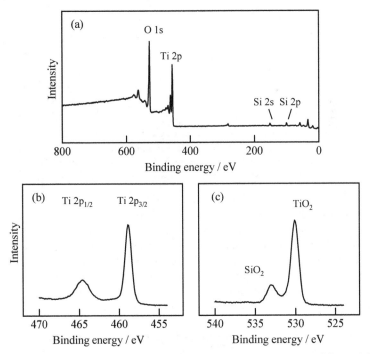

Fig. 6. XP spectra of hollow TiO$_2$ film on quartz glass substrate prepared from C-PS (0.28 %)
and 0.005 mol dm^{-3} TALH; (a) wide scan spectrum, (b) Ti 2p and (c) O 1s photoelectron
spectra.

On the hypothesis that the sample is constituted by isotropic and spherical crystallites, the width of the most intensive X-ray diffraction peak, corresponding to anatase (101) plane at 25.273°, is adopted for the calculation of the β value. With the β value: 0.0153 (radian) obtained after the exclusion of the effect of $K\alpha_2$ line and optical system of the instrument, the D value is estimated to be 7.7 nm.

Figure 6 shows XP spectra of the surface of hollow shells film. In the wide energy range spectrum (Fig. 6 (a)) small peaks of Si 2p and Si 2s photoelectron originating from quartz glass substrate were observed as well as main titanium and oxygen peaks. Almost identical peak positions (within 1 eV) of Ti $2p_{3/2}$ region spectrum for the film and standard TiO_2 (Chigane et al., 2009) at 458.9 and 458.7 eV, respectively, suggest that the chemical state of the hollow shells film can be assigned to TiO_2. The main peak and secondary peak at 530 eV and at around 533 eV in O 1s photoelectron region spectrum can be attributed to titanium oxide (Ti–O–Ti) and silica of substrate, respectively, from references. (Moulder et al., 1992)

Figure 7 shows optical properties of H-TiO_2 films prepared from C-PS and TALH.

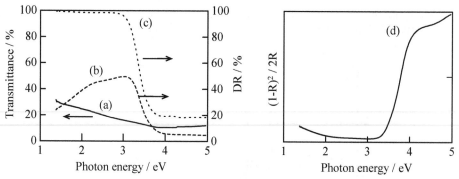

Fig. 7. Relationship between H-TiO_2 film on quartz glass substrate from C-PS-TALH and UV-visible photon energy: (a, solid line) transmission spectrum of the film and (b, dashed line) or (c, dotted line) diffuse reflection spectrum of film or powder sample, respectively and (d) Kubelka-Munk function plots for the powdered sample transformed from DR spectrum corresponding to (c).

Absence of clear attenuation dip in transmission spectrum indicates random configuration of hollow shells. Decrease and increase of transmittance and DR of film samples, respectively, according to photon energy up to ca. 3 eV are due to scattering of light by aggregation of TiO_2 hollows. Taking account of DR of powder sample the decrease of both transmittance and DR of film sample at higher than 3 eV are due to interband photo-absorption of TiO_2. Such change of DR of powder sample can be associated with absorption coefficient (α) using Kubelka-Munk function (Kortüm, 1969; Murphy, 2007)

$$F(R) = \frac{(1-R)^2}{2R} = \frac{\alpha}{S} \qquad (2)$$

as shown in Fig. 7(d), where R and S indicate diffuse reflectance and scattering coefficient, respectively. Although all terms depend on energy (wavelength) of incident light, the sizes of hollow shells collectives (more than 1 μm) are enough larger than wavelength and in the wavelength range between 310 nm (ca. 4 eV) and 410 nm (ca. 3 eV) concerning photoabsorption (Fig. 7(c)) the change of S value might be little. Therefore in relation to absorption S can be assumed to be constant and then we investigated band edge using α/S values. The relationship of α and photon energy (hv) of semiconductors near the absorption edge region for direct or indirect transition is given by

$$(\alpha h v)^2 = A(h v - E_g) \qquad (3)$$

or

$$(\alpha h v)^{1/2} = A(h v - E_g) \qquad (4)$$

respectively, where A and E_g are proportion constant and band gap energy.(Pankove, 1971) The linear relationship of Eq. (3) and (4) indicate that in the plots of $(\alpha h v)^2$ or $(\alpha h v)^{1/2}$ versus hv the optical band gap can be determined by intersection of extrapolated straight line with hv axis. (Barton et al., 1999; Han et al., 2007; Nowak et al., 2009) Based on above mentioned assumption we derived the optical band gaps as 3.5 eV and 3.2 eV from intersection points of linear fitting line with $(\alpha h v/S)^2$ and $(\alpha h v/S)^{1/2}$ versus hv for direct and indirect process, respectively, as shown in Fig. 8. At present we propose the latter value because 3.2 eV of E_g for anatase TiO_2 is commonly accepted.(O'Regan & Grätzel, 1991; Tang et al., 1994) However more investigation is necessary since it has not been clarified whether interband transition of anatase TiO_2 is direct or indirect. (Asahi et al., 2000; Mo & Ching, 1995)

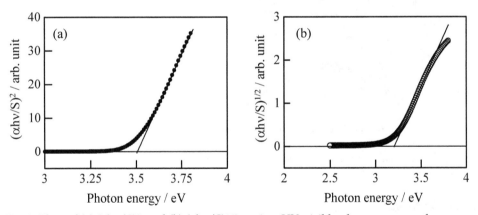

Fig. 8. Plots of (a) $(ahv/S)^2$ and (b) $(ahv/S)^{1/2}$ against UV-visible photon energy of powdered $H-TiO_2$ transferred from a / S spectrum corresponding to Fig. 7(d). Linear lines are drawn to determine band gap energies.

3.2 Enhanced film quality by electrodeposition

Figure 9(a) shows low magnification SEM image of the films indicating considerable volume change by calcination. After calcination the film was constricted and broken apart as expected in introduction section. Moreover maybe owing to the stress by crystallization and plastic strain the film easily detached from substrate. Such poor quality of the films and large crack made us expect low utility for DSSC electrode. Figure 9(b) and (c) show SEM images of TiO_2-coated H-TiO_2 film by electrodeposition before calcination. Although on the top surface view some narrow cracks were observed within 1 μm width, recognition of only hollow spheres in the back of the cracks indicates that the FTO substrate are not exposed. Cross-sectional view shows hollow shells suggesting successful maintenance of PS-TALH structure during electrodeposition.

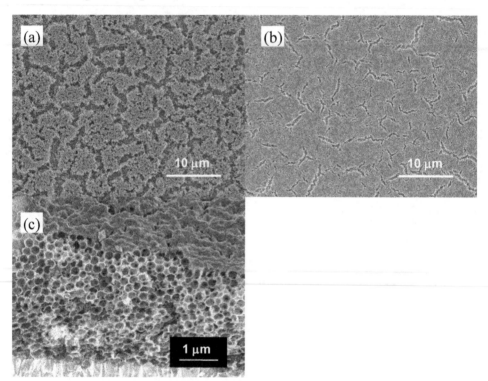

Fig. 9. SEM photographs of H-TiO_2 films on FTO substrates; (a) or (b) surface view of the film without or with electrodeposition TiO_x coating, respectively, and (c) cross-sectional view of the film with electrodeposition TiO_x coating. In both cases the precursors are C-PS-TALH.

3.3 DSSC properties
Figure 10 and Table 1 show typical results of DSSC assessment for three type TiO_2 electrodes. The conversion efficiency value of the cell using only H-TiO_2 electrode: 0.91 % (Fig. 10(a)) was about 4 times lower than that for the cell: 3.98 % using standard nc-TiO_2 electrode (Fig. 10(b)).

The conversion efficiency value of the cell using only H-TiO_2 electrode: 0.91 % (Fig. 10(a)) was about 4 times lower than that for the cell: 3.98 % using standard nc-TiO_2 electrode (Fig. 10(b)), despite 3.2 times smaller amount of TiO_2. There can be thought to be two reasons for such insufficient efficiency. One is lower short circuit current density (J_{SC}). The compli-cated morphology of hollow TiO_2 film might contribute to the increase of specific contact area with dye molecules and to harvesting light by scattering in the hollow shells. However a large number of broken points in the skeleton structure and few contact points with substrate supposedly leaded poor conduction of photo-induced electrons. The other is lower open circuit voltage (V_{OC}) due to direct contact between electrolyte solution and FTO substrate through wider film crack points. So as to improve the two problems we inserted initial flat TiO_2 layer on the substrate by electrolysis of TALH solution. Remarkably improved DSSC performance: 2.90 % for the composite (C-TiO_2) film composed of electrodeposited TiO_2 bottom layer and hollow TiO_2 top layer has been shown (Fig. 10(c) and Table 1) compared with single hollow layer (0.91 %), including enhancement of V_{OC} and FF values. The area density of TiO_2 of nanoparticles electrode was as more than 2.4 times as the composite hollow shell film, whereas the change of the conversion efficiency was 1.4 times. This indicates the conversion efficiency per-TiO_2-weight for the composite hollow shell film is higher than nc-TiO_2 film owing to homogeneity effect of blocking layer making complicated structure of H-TiO_2 available

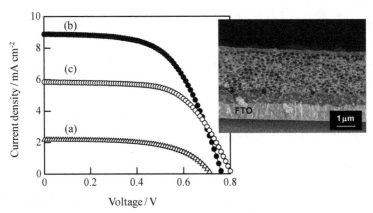

Fig. 10. Photocurrent density voltage curves of DSSCs using (a) H-TiO_2, (b) nc-TiO_2 and (b) C-TiO_2 film. The cross-sectional SEM photograph indicates the C-TiO_2 sample corresponding to curve (c).

TiO_2 films	Area density of TiO_2 / mg cm^{-2} (by XRF)	J_{SC} / mA cm^{-2} [a]	V_{OC} / V [b]	FF[c]	Eff / % [d]
Hollow shell	0.179	2.19	0.716	0.580	0.911
Composite	0.244	5.85	0.805	0.615	2.90
Nano-particles	0.587	8.89	0.763	0.585	3.98

[a]Short circuit current density. [b]Open circuit voltage. [c]Fill factor. [d]Conversion efficiency.

Table 1. DSSC properties of two TiO_2 film electrodes.

Figure 11 shows IPCE assessments of cells using three type TiO_2 electrodes: nc-TiO_2, simply electrodeposited TiO_2 (E-TiO_2) and C-TiO_2 film. In this comparison we electrodeposited thicker first film than that in above J-V characterization. By this effect the conversion efficiency of C-TiO_2 film was somewhat enhanced to 3.44 %. The normalized IPCE curves for the films (Fig. 11(e), (f), (g)) have shown good spectral accordance of photocurrent with photoabsorption of dye (Fig. 10(d)). Moreover heightening of composite TiO_2 film (Fig. 11(f)) by 14 % and 11 % at 450 nm and 600 nm of wavelength, respectively, against simple electrodeposited film (Fig. 11(e)), proves wavelength-independent increase of photocurrent by the addition of hollow shells as top layer. This seems to arise simply from increase of amount of TiO_2 and to imply low optical effect of our hollow shells, remaining what should be improved.

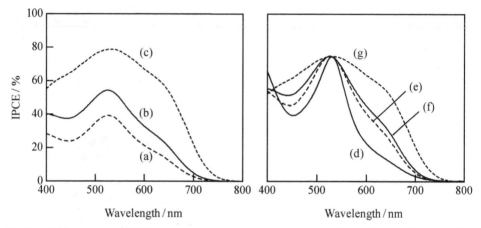

Fig. 11. IPCE spectra of DSSC using (a, dashed line) or (e, dashed line) E-TiO_2 film, (b, solid line) or (f, solid line) C-TiO_2 film and (c, dotted line) or (g, dotted line) nc-TiO_2 film.

4. Conclusions

Hollow TiO_2 (H-TiO_2) shell films have been prepared using environmentally benign Ti-lactate complex (TALH) as a titanium source. A precursor has been prepared from an aqueous colloidal suspension containing both polystyrene (PS) and TALH. Successive calcination of the PS-TALH precursor at 723 K led to the decomposition of PS and formation of hollow spherical shells. Employing PS with cationic surface functional group (C-PS) gave rise to smoothly spreading of hollow shells films in wider area compared with conventional PS possessing anionic groups (A-PS). The characterizations by X-ray diffraction and X-ray photoelectron spectroscopy have proved that the hollow shell films can be assigned to anatase TiO_2. Optical characteristics of H-TiO_2 films from C-PS-TALH have shown they are not in an inverse opal structure. Diffuse reflection spectroscopy combined with Kubelka-Munk treatment has revealed wide optical band gap as equal or more than 3.2 eV, indicating transparency of the films for visible light. It has been found that combination with electrodeposition of TiO_x films strongly supports quality and DSSC properties of H-TiO_2 films in two aspects. First the electrodeposition onto the PS-TALH precursor film effectively prevented widely cracking in the calcination process. Moreover the electrodeposition onto FTO substrate before H-TiO_2 layer, namely fabrication of composite film (C-TiO_2 film), substantially enhanced DSSC energy conversion efficiency: 2.9 % which was comparable to an electrode using commercially available TiO_2 nanocrystalline particle (nc-TiO_2). The IPCE curve of C-TiO_2 film has revealed the increase of photocurrent due to light-harvesting effect.

On the other hand its independence of wavelength has indicated that H-TiO$_2$ lacks photonic crystal effect and then a future investigation on improvement of arraying of H-TiO$_2$ is needed.

5. References

Asahi, R.; Taga, Y.; Mannstadt, W. & Freeman, A. J. (2000). Electronic and optical properties of anatase TiO$_2$, *Physical Review B*, Vol.61, No.11, pp 7459–7465, ISSN 1098-0121

Barton, D. G.; Shtein, M.; Wilson, R. D.; Soled, S. L. & Iglesia, E. (1999). Structure and Electronic Properties of Solid Acids Based on Tungsten Oxide Nanostructures, *Journal of Physical Chemistry B*, Vol.103, No.4, pp 630-640, ISSN 1520-6106

Bisquert, J.; Cahen, D.; Hodes, G.; Rühle, S. & Zaban, A. (2004). Physical Chemical Principles of Photovoltaic Conversion with Nanoparticulate, Mesoporous Dye-Sensitized Solar Cells, *Journal of Physical Chemistry B*, Vol.108, No.24, pp 8106–8118, ISSN 1520-6106

Caruso, F.; Shi, X.; Caruso, R. A. & Susha, A. (2001). Hollow Titania Spheres from Layered Precursor Deposition on Sacrificial Colloidal Core Particles, *Advanced Materials*, Vol.13, No.10, pp 740–744, ISSN 0935-9648

Chigane, M.; Watanabe, M.; Izaki, M.; Yamaguchi, I. & Shinagawa, T. (2009). Preparation of Hollow Titanium Dioxide Shell Thin Films by Electrophoresis and Electrolysis for Dye-Sensitized Solar Cells, *Electrochemical and Solid-State Letters*, Vol.12, No.5, pp E5–E8, ISSN 1099-0062

Enright, B. & Fizmaurice, D. (1996). Spectroscopic Determination of Electron and Hole Effective Masses in a Nanocrystalline Semiconductor Film, *Journal of Physical Chemistry*, Vol.100, No.3, pp 1027–1035, ISSN 0022-3654

Galusha, J. W.; Tsung, C-K.; Stucky, G. D. & Bartl, M. H. (2008). Optimizing Sol-Gel Infiltration and Processing Methods for the Fabrication of High-Quality Planar Titania Inverse Opals, *Chemistry of Materials*, Vol.20, No.15 pp 4925–4930, ISSN 0897-4756

Guldin, S.; Httner, S.; Kolle, M.; Welland, M. E.; Müller-Buschbaum, P.; Friend, R. H.; Steiner, U. & Ttreault, N. (2010). Dye-Sensitized Solar Cell Based on a Three-Dimensional Photonic Crystal, *Nono Letters*, Vol.10, No.7, pp 2303–2309, ISSN 1530-6984

Han, T. -Y.; Wu, C. -F. & Hsieh, C. -T. (2007). Hydrothermal synthesis and visible light photocatalysis of metal-doped titania nanoparticles, *Journal of Vacuum Science & Technology, B: Microelectronics and Nanometer Structures--Processing, Measurement, and Phenomena*, Vol.25, No.2, pp 430-435, ISSN 1071-1023

Hartsuiker, A. & Vos, W. L. (2008). Structural Properties of Opals Grown with Vertical Controlled Drying, *Langmuir*, Vol.24, No.9, pp 4670–4675, ISSN 0743-7463

International Center for Diffraction Data, Joint Committee on Powder Diffraction Standards, (1990). *Powder Diffraction File*, set 21, No.1272, Swarthmore, Pennsylvania, USA

Kang, S. H.; Choi, S-H.; Kang, M-S.; Kim, J-Y.; Kim, H-S.; Hyeon, T. & Sung, Y. -E. (2008). Nanorod-Based Dye-Sensitized Solar Cells with Improved Charge Collection Efficiency, *Advanced Materials*, Vol.20, No.1, pp 54–58, ISSN 0935-9648

Kang, T-S.; Smith, A. P.; Taylor, B. E. and Durstock, M. F. (2009). Fabrication of Highly-Ordered TiO$_2$ Nanotube Arrays and Their Use in Dye-Sensitized Solar Cells, *Nono Letters*, Vol.9, No.2, pp 601–606, ISSN 1530-6984

Karuppuchamy, S.; Amalnerkar, D. P.; Yoshida, T.; Sugiura, T. & Minoura, H. (2001). Cathodic electrodeposition of TiO$_2$ thin films for dye-sensitized photoelectrochemical applications, *Chemistry Letters*, Vol.30, No.1, pp 78-79, ISSN 0366-7022

Kim, G. -M.; Lee, S. -M.; Michler, G. H.; Roggendorf, H.; Gösele, U. & Knez, M. (2008). Nanostructured pure anatase titania tubes replicated from electrospun polymer fiber templates by atomic layer deposition, *Chemistry of Materials*, Vol.20, No.9, pp 3085-3091, ISSN 0897-4756

King, J. S.; Grangnard, E. & Summers, C. J. (2005). TiO$_2$ Inverse Opals Fabricated Using Low-Temperature Atomic Layer Deposition, *Advanced Materials*, Vol.17, No.8, pp 1010–1013, ISSN 0935-9648

Kondo, Y.; Yoshikawa, H.; Awaga, K.; Murayama, M.; Mori, T.; Sunada, K.; Bandow, S. & Iijima S. (2008). Preparation, Photocatalytic Activities, and Dye-Sensitized Solar-Cell Performance of Submicron-Scale TiO$_2$ Hollow Spheres, *Langmuir*, Vol.24, No.2, pp 547–550, ISSN 0743-7463

Kortüm, G. (1969). *Reflectance Spectroscopy*, pp 103-127, Springer-Verlag, ISBN 978-3540045872, New York, USA

Kwak, E. S.; Lee, W.; Park, N-G.; Kim, J. & Lee, H. (2009). Compact inverse-opal electrode using non-aggregated TiO$_2$ nanoparticles for dye-sensitized solar cells, *Advanced Functional Materials*, Vol.19, No.7, pp 1093–1099, ISSN 1616-301X

Liu, W.; Zou, B.; Zhao, J. & Cui, H. (2010). Optimizing sol-gel infiltration for the fabrication of high-quality titania inverse opal and its photocatalytic activity, *Thin Solid Films*, Vol.518, No.17, pp 4923–4927, ISSN 0040-6090

Mo, S. -D. & Ching, W. Y. (1995). Electronic and optical properties three phases of titanium dioxide: Rutile, anatase, and brookite, *Physical Review B*, Vol.51, No.19, pp 13023–13032, ISSN 1098-0121

Moulder, J. F.; Stickle, W. F.; Sobol P. E. & Bomben, K. D. (1992). *Handbook of X-ray Photoelectron Spectroscopy*, Perkin-Elmer, p. 45, ISBN 0-9627026-2-5, Minnesota, USA

Murphy, A. B. (2007). Band-gap determination from diffuse reflectance measurements of semiconductor films, and application to photoelectrochemical water-splitting, *Solar Energy Materials and Solar Cells*, Vol.91, No.14, pp 1326-1337, ISSN 0927-0248

Nazeeruddin, M. K.; Péchy, P.; Renouard, T.; Zakeeruddin, S. M.; Humphry-Baker, R.; Comte, P.; Liska, P.; Cevey, L.; Costa, E.; Shklover, V.; Spiccia, L.; Deacon, G. B.; Bignozzi, C. A. & Grätzel, M. (2001). Engineering of Efficient Panchromatic Sensitizers for Nanocrystalline TiO$_2$-Based Solar Cells, *Journal of the American Chemical Society*, Vol.123, No.8, pp 1613–1624, ISSN 0002-7863

Nishimura, S.; Shishido, A.; Abrams, N. & Mallouk, T. E. (2002). Fabrication technique for filling-factor tunable titanium dioxide colloidal crystal replicas, *Applied Physics Letters*, Vol.81, No.24, pp 4532–4534, , ISSN 0003-6951

Nishimura, S.; Abrams, N.; Lewis, B. A.; Halaoui, L. I.; Mallouk, T. E.; Benkstein, K. D.; van de Lagemaat, J. & Frank, A. J. (2003). Standing Wave Enhancement of Red

Absorbance and Photocurrent in Dye-Sensitized Titanium Dioxide Photoelectrodes Coupled to Photonic Crystals, *Journal of the American Chemical Society*, Vol.125, No.20, pp 6306–6310, ISSN 0002-7863

Nowak, M.; Kauch, B. & Szperlich, P. (2009). Determination of energy band gap of nanocrystalline SbSI using diffuse reflectance spectroscopy, *Review of Scientific Instruments*, Vol.80, No.4, pp 046107-1-046107-3, ISSN 0034-6748

O'Regan, B. & Grätzel, M. (1991). A low-cost, high-efficiency solar cell based on dye-sensitized colloidal TiO_2 films, *Nature*, Vol.353, No.6346, pp. 737–740, ISSN 0028-0826

Pankove, J. I. (1971). *Optical processes in semiconductors*, pp 34-42, Dover Publications, ISBN 0-486-60275-3, New York, USA

Paulose, M.; Shankar, K.; Varghese, O. K.; Mor, G. K. & Grimes, C. A. (2006). Application of highly-ordered TiO_2 nanotube-arrays in heterojunction dye-sensitized solar cells, *Journal of Physics D: Applied Physics*, Vol.39, No.12, pp 2498–2503, ISSN 0022-3727

Peng, T. Y.; Hasegawa, A.; Qui, J. R. & Hirao, K. (2003). Fabrication of Titania Tubules with High Surface Area and Well-Developed Mesostructural Walls by Surfactant-Mediated Templating Method, *Chemistry of Materials*, Vol.15, No.10, pp 2011–2016, ISSN 0897-4756

Qi, L.; Sorge, J. D.; & Birnie III, D. P. (2009). Dye-sensitized solar cells based on TiO_2 coatings with dual size-scale porosity, *Journal of the American Ceramic Society*, Vol.92, No.9, pp 1921–1925, ISSN 0002-7820

Rouse, J. H. & Ferguson, G. S. (2002). Stepwise Formation of Ultrathin Films of a Titanium (Hydr) Oxide by Polyelectrolyte-Assisted Adsorption, *Advanced Materials*, Vol.14, No.2, pp 151–154, ISSN 0935-9648

Ruani, G.; Ancora, C.; Corticelli, F.; Dionigi, C. & Rossi, C. (2008). Single-step preparation of inverse opal titania films by the doctor blade technique, *Solar Energy Materials and Solar Cells*, Vol.92, No.5, pp 537–542, ISSN 0927-0248.

Tachibana, Y.; Moser, J. E.; Grätzel, M.; Klug, D. R. & Durrant, J. R. (1996). Subpicosecond Interfacial Charge Separation in Dye-Sensitized Nanocrystalline Titanium Dioxide Films, *Journal of Physical Chemistry*, Vol.100, No.51, pp 20056–20062, ISSN 0022-3654

Tang, H.; Prasad, K.; Sanjinès, R.; Schmid, P. E. & Lévy, F. (1994). Electrical and optical properties of TiO_2 anatase thin films, *Journal of Applied Physics*, Vol.75, No.4, pp 2042-2047, ISSN 0021-8979

Watanabe, M.; Aritomo, H.; Yamaguchi, I.; Shinagawa, T.; Tamai, T.; Tasaka, A. & Izaki, M. (2007). Selective Preparation of Zinc Oxide Nanostructures by Electrodeposition on the Templates of Surface-functionalized Polymer Particles, *Chemistry Letters*, Vol.36, No.5, pp 680–681, ISSN 0366-7022

Yamaguchi, K.; Sawatani, S.; Yoshida, T.; Ohya, T.; Ban, T.; Takahashi Y. & Minoura, H. (2005). Electrodeposition of TiO_2 thin films by anodic formation of titanate/benzoquinone hybrid, *Electrochemical and Solid-State Letters*, Vol.8, No.5, pp C69-C71, ISSN 1099-0062

Yang, S-C.; Yang, D-J.; Kim, J.; Hong, J-M.; Kim, H-G.; Kim, I-D. & Lee, H. (2008). Hollow TiO_2 Hemispheres Obtained by Colloidal Templating for Application in Dye-

Sensitized Solar Cells, *Advanced Materials*, Vol.20, No.5, pp 1059–1064, ISSN 0935-9648

Ye, Y. -H.; LeBlanc, F.; Hache, A.; Truong, V. -V. (2001). Self-assembling three-dimensional colloidal photonic crystal structure with high crystalline quality, *Applied Physics Letters*, Vol.78, No.1, pp 52-54, ISSN 0003-6951

Yip, C. -H.; Chiang, Y. -M. & Wong, C. -C. (2008). Dielectric Band Edge Enhancement of Energy Conversion Efficiency in Photonic Crystal Dye-Sensitized Solar Cell, *Journal of Physical Chemistry C*, Vol.112, No.23, pp 8735–8740, ISSN 1932-7447

8

Carbon Nanostructures as Low Cost Counter Electrode for Dye-Sensitized Solar Cells

Qiquan Qiao
South Dakota State University
United States

1. Introduction

In the last two decades, dye sensitized solar cells (DSSCs) have gained extensive attention as a low cost alternative to conventional Si solar cells (Oregan & Gratzel 1991; Fan et al. 2008; Xie et al. 2009; Alibabaei et al. 2010; Gajjela et al. 2010; Xie et al. 2010; Yum et al. 2010). A typical DSSC is made of a TiO_2 photoanode and a Pt counter electrode separated by an electrolyte comprising an iodide/triiodide ($I^-/I3^-$) redox couple. The photoanode is usually prepared from TiO_2 nanoparticles on a transparent conducting oxide (TCO), while the counter electrode is a thin layer of Pt deposited on another TCO substrate. The dye molecules are adsorbed onto TiO_2 surface. When exposed to sunlight, photoelectrons are generated and injected into the photoanode. Afterward, the electrons travel to counter electrode through an outside load. The oxidized dye molecules then retake electrons from I^- ions and oxidize I^- into $I3^-$. Meanwhile, the $I3^-$ is reduced into I^- by taking electrons from counter electrode. Pt counter electrode has been extensively used as an efficient electrocatalyst for reduction of $I3^-$ ions in DSSCs (Gratzel 2003; Sun et al. 2010). However, Pt is an expensive metal and can also be corroded by $I^-/I3^-$ redox couple (Kay & Gratzel 1996). Recently, various carbonaceous materials including graphite, carbon black, and carbon nanotubes have been studied as a low cost replacement for Pt as an electrocatalyst for reduction of $I3^-$ ions (Kay & Gratzel 1996; Burnside et al. 2000; Imoto et al. 2003; Imoto et al. 2003; Suzuki et al. 2003; Murakami et al. 2006; Ramasamy et al. 2007; Fan et al. 2008; Hinsch et al. 2008; Joshi et al. 2009; Lee et al. 2009; Skupien et al. 2009; Calandra et al. 2010). The carbonaceous materials are plentiful, inexpensive, and also exhibit high resistivity to corrosion (Ramasamy et al. 2007). Replacement of Pt with carbon-based materials can also speed up DSSC commercialization (Burnside et al. 2000; Hinsch et al. 2008; Han et al. 2009; Skupien et al. 2009; Joshi et al. 2010).

In this chapter, we review some carbon nanostructures including carbon nanoparticles and electrospun carbon nanofibers that have been successfully used as a low cost alternative to Pt in DSSCs. The carbon nanoparticle- and carbon nanofiber-based DSSCs showed comparable performance as that of Pt-based devices in terms of short circuit current density (Jsc) and open circuit voltage (Voc). Electrochemical impedance spectroscopy (EIS) measurements indicated that the carbon nanoparticle and carbon nanofiber counter electrodes showed lower charge transfer resistance (R_{ct}), suggesting that carbon nanoparticle and carbon nanofiber counter electrodes are an efficient electrocatalyst for DSSCs. In addition, the series resistance of carbon-based counter electrodes was found to be a little higher than that of Pt cells, leading to a slightly lower FF. Herein, we will first introduce the preparation and characterization of carbon nanoparticle and carbon nanofiber counter electrodes. Then, the fabrication of DSSC devices with these carbon-based counter electrodes will be described and compared with Pt-based cells. The use of carbon nanoparticle and carbon nanofiber counter electrodes has a great potential to make low cost DSSC technology one step closer to commercialization.

2. Carbon/TiO₂ composite as counter electrode

Low cost carbon/TiO_2 composite was used as an alternative to platinum as a counter-electrode catalyst for tri-iodide reduction. In the carbon/TiO_2 composite, carbon is nanoparticles and acts as an electrocatalyst for triiodide reduction, while the TiO_2 functions as a binder. The carbon/TiO_2 composite can be deposited by spin coating or doctor blading onto a fluorine-doped Tin Dioxide (FTO).

2.1 Preparation of carbon/TiO₂

Carbon nanoparticles (Sigma-Aldrich) have a particle size < 50 nm and a surface area > 100 m^2/g. The TiO_2 paste was prepared by dispersing TiO_2 nanoparticles (P25 Degussa, average size of 25 nm) into water. The carbon/TiO_2 composite was made by mixing 650 mg carbon nanoparticles with 1 ml TiO_2 colloid paste at a concentration of 20 wt%. Then 2 ml deionized (DI) water was added, followed by grinding and sonication. 1 ml Triton X-100 was added during grinding. The final paste was then spin coated onto a FTO glasses to form the counter electrode, followed by sintering at 250⁰ C for an hour.

The scanning electron microscopy (SEM) images of carbon/TiO_2 composite and pure TiO_2 nanoparticle films are shown in Figure 1a and b, respectively. It can be seen that the carbon/TiO_2 composite counter-electrode film is highly porous with a large surface area, which can function effectively for tri-iodide reduction. The pore size ranges from 20 nm to 200 nm throughout the film, which is large enough for I^-/I_3^- ions that are only a few angstroms to diffuse into the pores and get reduced at the carbon nanoparticle surface(Ramasamy et al. 2007). The particle size in carbon/TiO_2 composite film (Figure 1a) is apparently larger than those in pure TiO_2 nanoparticle film (Figure 1b). This suggests that the carbon nanoparticle dominates in carbon/TiO_2 mixture and effectively serves as a catalyst for tri-iodide reduction. A cross-section SEM image (Figure 1c) shows that the carbon/TiO_2 composite counter electrode has a thickness of about 11.2 um.

(a)

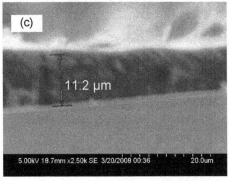

Fig. 1. SEM images of (a) 11.2 um thick carbon/TiO₂ composite layer and (b) pure TiO₂ nanoparticle layer on a FTO substrate. Cross section SEM image of (c) the carbon/TiO₂ composite layer. Reproduced with permission from Ref (Joshi et al. 2009).

2.2 Calculation of series resistance, left justified

Ramasamy et al. measured the charge transfer resistance (R_{ct}) of carbon electrode via electrochemical impedance spectroscopy (EIS) and found that R_{ct} was 0.74 Ω cm^{-2}, two times less than that of the screen printed Pt (Ramasamy et al. 2007). Since the thickness of carbon-based counter electrode is tens of micrometers that are much higher than Pt at a thickness of about tens of nanometers, the internal series resistance (R_{se}) of carbon-based DSSCs are found to be higher (Ramasamy et al. 2007; Joshi et al. 2009). The lower R_{ct} counterbalances the higher R_{se} of carbon-based device. The series resistance of carbon/TiO₂ composite based DSSCs was also studied and compared with that of platinum-based devices under multiple light intensities.

Current density (J_{sc}) through the series resistance is as below (Matsubara et al. 2005):

$$J = J_{PH} - \frac{V - IR_s}{AR_{sh}} - J_0\left(\exp[q(V - JA\ R_s)/nkT] - 1\right) \tag{1}$$

This equation can be modified as:

$$J_{PH} - J = \frac{V - JAR_s}{AR_{sh}} + J_0\left(\exp[q(V - JA\ R_s)/nkT] - 1\right) \tag{2}$$

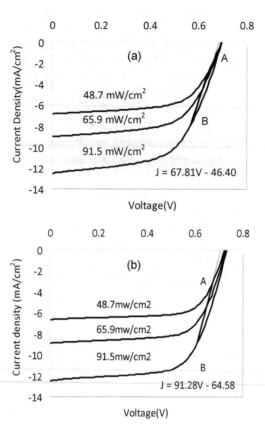

Fig. 2. J-V curves of DSSC devices at different light intensity from (a) carbon/TiO$_2$ composite and (b) Pt counter electrode. Reproduced with permission from Ref (Joshi et al. 2009).

When we plot current density-voltage (J-V) curves at multiple light intensities and select the points of (J,V) which satisfy the following condition:

$$J_{PH} - J = \Delta J = constant \qquad (3)$$

The points should lie in the straight line and follow:

$$J = V / RsA + constant \qquad (4)$$

Thus, the series resistance can be determined from the slope of a straight line. The current density-voltage (J-V) curves at different light intensities of the carbon/TiO$_2$-based and Pt-based DSSC devices are shown in Figure 2a and b, respectively.

2.3 Device performance of carbon/TiO$_2$ composite counter electrode

The active area of carbon/TiO$_2$ composite is 0.20 cm^2, while that of Pt devices is 0.24 cm^2. The slope of the straight line AB in carbon/TiO$_2$ composite devices is 67.81 mA/(cm^{2}V), with a reciprocal of 14.75 Ωcm^2 . The slope of the straight line AB in Pt-based devices is 91.28 mA/(cm²V) and its reciprocal is 11.37 Ωcm^2. Apparently the series resistance of carbon/TiO_2 devices is larger than that of Pt devices. This can be possibly attributed to the much thicker layer and larger resistivity of carbon/TiO_2 counter electrode than those of Pt (Imoto et al. 2003). However, the carbon/TiO_2 counter electrode has its own advantage that is the large surface area. This results in a lower R_{ct}, which was found to be less than half of that in the Pt counter electrode (Ramasamy et al. 2007). The lower R_{ct} can compensate the effects of higher series resistance.

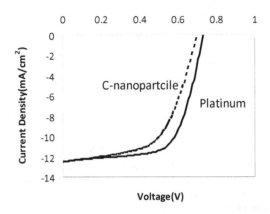

Fig. 3. J-V curves of DSSC devices with carbon/TiO_2 composite (dash line) and Pt (solid line) counter electrode under AM 1.5 illumination (light intensity: 91.5 mW/cm²). Reproduced with permission from Ref (Joshi et al. 2009).

Figure 3 shows a comparison of J-V curves from carbon/TiO_2 and Pt devices under an AM 1.5 solar simulator at an intensity of ~ 91.5 mW/cm². DSSCs with carbon/TiO_2 counter electrode achieve an efficiency of 5.5 %, which is comparable to 6.4 % of Pt counter electrode devices. The photovoltaic parameters in terms of short circuit current density (Jsc), open circuit voltage (Voc), fill factor (FF) and efficiency (η) are listed in Table 1. The FF of carbon/TiO_2 devices was found to be slightly lower than Pt devices. This may be attributed to higher series resistance (14.75 Ωcm^2) in the former compared to that (11.37 Ωcm^2) in the latter. Ramasamy et al. studied the robustness of carbon-based DSSCs and their results showed that carbon-based cells have a comparable stability as Pt-based devices (Ramasamy et al. 2007).

Counter electrodes	Jsc (mA/cm²)	Voc (V)	FF	η	Rs (Ω)
carbon/TiO_2 composite	12.53	0.70	0.57	5.5 %	14.75 Ωcm^2
Platinum	12.48	0.73	0.65	6.4 %	11.37 Ωcm^2

Table 1. DSSC device parameters from carbon/TiO_2 composite and Pt counter electrode. Reproduced with permission from Ref (Joshi et al. 2009).

3. Carbon nanofibers as counter electrode

Carbon nanofibers prepared by electrospinning were also explored as low cost alternative to Pt for triiodide reduction catalyst in DSSCs. The carbon nanofiber counter electrode was characterized by EIS and cyclic voltammetry measurements. The carbon nanofiber counter electrode exhibited low charge transfer resistance (R_{ct}), small constant phase element (CPE) exponent (β), large capacitance (C), and fast reaction rates for triiodide reduction.

3.1 Preparation of carbon nanofiber counter electrode

The carbon nanofiber paste was made by mixing 0.1 g ECNs with 19.6 g polyoxyethylene(12) tridecyl ether (POETE) in a similar method reported by others (Mei & Ouyang 2009). The mixture was then grinded, sonicated, and centrifuged at a spin speed of 10,000 rpm to uniformly disperse the ECNs in POETE. Any extra POETE that floated on top of the mixture after the centrifuge was removed via a pipette. Afterwards, the counter electrode was made by doctor-blading the mixture onto FTO (~8 Ω/$\square$ and ~400 nm), followed by sintering at 200 °C for 15 min and then at 475 °C for 10 min. Figure 4 shows SEM and transmission electron microscope (TEM) images of the original carbon nanofibers prepared by electrospinning and the carbon nanofiber counter electrode on FTO deposited by doctor blading. In the original electrospun carbon nanofiber samples, the ECNs were relatively uniform in diameter with an average value of ~ 250 nm (Figure 4a). The TEM image in Figure 4b shows that the structure of ECNs was primarily turbostratic instead of graphitic; *i.e.*, tiny graphite crystallites with sizes of a few nanometers were embedded in amorphous carbonaceous matrix. The nanofiber sheet did not show evidence of microscopically identifiable beads or beaded-nanofibers. The BET surface area of the carbon nanofiber sheet was measured to be ~100 m^2/g via a Micromeritics ASAP 2010 surface area analyzer using N$_2$ adsorption at 77 K.

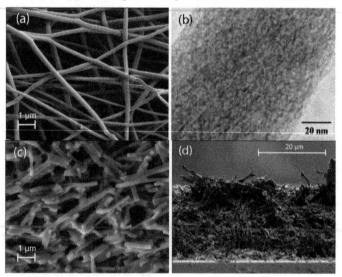

Fig. 4. (a) SEM image of electrospun carbon nanofiber film; (b) TEM image of a typical single carbon nanofiber; SEM image of (c) top-view and (d) cross-section of carbon nanofiber counter electrode. Reprinted with permission from {Joshi et al. 2010}. Copyright {2010} American Chemical Society.

Because it was difficult to attach the original carbon nanofiber sheet onto FTO, we added POETE into the carbon nanofiber, followed by grinding and sonication. As shown in Figure 4c, the nanofibers that were originally tens of microns long were broken into submicrons to microns after grinding and sonication. The conductivity of original electrospun carbon nanofibers (Figure 4a) is ~ 1538 Sm^{-1}, but decreased to ~ 164 Sm^{-1} after converted to the counter electrode as shown in Figure 4c. This can possibly be attributed to the much smaller lengths of the carbon nanofibers that reduced conduction network. Also, the POETE was burned away at high temperature, causing additional voids between carbon nanofibers. However the smaller length of carbon nanofibers may increase the surface area of the counter electrode, which can be seen by comparing Figure 4a with Figure 4c. The thickness of counter electrode was about of 24 µm (Figure 4d), which is much higher than that of carbon nanoparticle counter electrodes. The effects of carbon nanoparticle counter electrode thickness on DSSC parameters including Jsc, Voc, FF and cell efficiency (η) was studied by others (Murakami et al. 2006). They found that the thickness mainly affects FF and the optimal thickness was ~ 14.5 µm for carbon nanoparticle counter electrode. A thickness of ~11.2 µm was used in a carbon nanoparticle counter electrode DSSC device (Joshi et al. 2009). However, Ramasamy et al. prepared a carbon nanoparticle counter electrode with a larger thickness of ~ 20 µm (Ramasamy et al. 2007). Here, the thickness of carbon nanofiber counter electrode was higher than that of typical carbon nanoparticle counter electrode. As shown in Figure 4c, the shorter nanofibers are loosely packed with large voids and this can lead to smaller surface area than that of carbon nanoparticle counter electrode. A higher thickness was used to make the carbon nanofiber counter electrode to ensure a significant surface area.

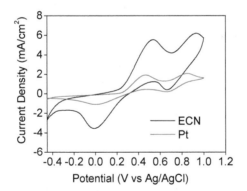

Fig. 5. Cyclic voltammograms of carbon nanofiber (black) and Pt (red) counter electrode. The measurement was performed in an acetonitrile solution comprising 10 mM LiI and 0.5 mM I$_2$. 0.1M tetra-n-butylammonium tetrafluoroborate was used as supporting electrolyte. Ag/ AgCl was used as reference electrode. The thickness of carbon nanofiber and Pt counter electrode is ~24 µm and ~40 nm, respectively. Reprinted with permission from {Joshi et al. 2010}. Copyright {2010} American Chemical Society.

3.2 Characterization of carbon nanofiber counter electrode
Cyclic voltammograms (Figure 5) of the carbon nanofiber and Pt counter electrode were performed in an acetonitrile solution that comprises 10 mM LiI and 0.5 mM I$_2$ using 0.1 M

tetra-n-butylammonium tetrafluoroborate as the supporting electrolyte. In the cyclic voltammetry (CV) measurements, Pt wire was used as counter electrode, Ag/AgCl as reference electrode, and a carbon nanofiber or Pt coated FTO as working electrode. Two pairs of oxidation and reduction peaks were found that are similar to those in the Pt electrodes. The oxidation and reduction pair on the left was from the redox reaction of $I_3^- + 2e^- = 3I^-$, while that on the right was attributed to the redox reaction of $3I_2 + 2e^- = 2I_3^-$ (Sun et al. ; Huang et al. 2007). The right pair from the carbon nanofiber sample exhibited a larger oxidation current density, but a smaller reduction current density than those of Pt electrode. This pair that was assigned to $3I_2 + 2e^- = 2I_3^-$ had little effect on DSSC performance (Mei et al. 2010). The left pair of carbon nanofiber counter electrode showed both a larger oxidation and reduction current density than those of the Pt electrodes. This pair that was assigned to $I_3^- + 2e^- = 3I^-$ directly affected DSC performance, indicating a fast rate of triiodide reduction.

The catalytic properties of counter electrode are usually characterized by EIS (Papageorgiou et al. 1997; Hauch & Georg 2001). In order to eliminate the effects of TiO₂ photoanode, a symmetrical carbon nanofiber – carbon nanofiber and Pt-Pt cells were fabricated for EIS study. These cells were prepared by assembling two identical carbon nanofiber (or Pt) electrodes face to face that were separated with an electrolyte of I^-/I_3^- redox couple. The EIS characterization was performed using an Ametek VERSASTAT3-200 Potentiostat equipped with frequency analysis module (FDA). The amplitude of AC signal was 10 mV with a frequency range of 0.1 - 10^5 Hz. The Nyquist plots of the symmetrical carbon nanofiber – carbon nanofiber and Pt-Pt cells are shown in Figure 6. Figure 6b shows the equivalent circuit that was used to fit impedance spectra. The equivalent circuit included charge transfer resistance (R_{ct}) at the carbon nanofiber or Pt electrode/electrolyte interface, constant phase element (CPE), series resistance (R_s) and Warburg impedance (Z_W) (Murakami et al. 2006). The R_{ct} at the electrode/electrolyte interface can be obtained from the high frequency semicircle, while the Z_W of the I^-/I_3^- redox couple in the electrolyte can be fitted from the low frequency arc (Wang et al. 2009; Jiang et al. 2010; Li et al. 2010; Mei et al. 2010). The fitted results from the Nyquist plots were summarized in Table 2. The R_{ct} of carbon nanofiber counter electrode was 0.7 Ωcm², less than half of that (1.9 Ωcm²) of the Pt electrode, suggesting a sufficient electro-catalytic capability. The CPE represents the capacitance at the interface between the carbon nanofiber or Pt and electrolyte, which can be described as:

$$Z_{CPE} = \frac{1}{Y_0}(j\omega)^{-\beta} \tag{5}$$

in which Y_0 is the CPE parameter, ω the angular frequency, and β the CPE exponent ($0 < \beta < 1$), and. The Y_0 and β are constant that is independent of frequency.

An ideal capacitance has a perfect semicircle where β is equal to 1. However, the porous films, leaky capacitor, surface roughness and non-uniform current distribution frequently cause a non-ideal capacitance that deviates β value away from 1 (Hauch & Georg 2001; Murakami et al. 2006). The fitted β value of the carbon nanofiber counter electrode was 0.82, smaller than that (0.95) of the Pt electrode. A lower β value suggested a higher porosity in carbon nanofiber electrode than that of Pt electrode (Murakami et al. 2006). In previous study, a β value of 0.81 was found in a highly porous carbon nanoparticle counter electrode (Murakami et al. 2006). Also, the capacitance (C) in carbon nanofiber counter electrode was larger than that of Pt electrode, suggesting a higher surface area in carbon nanofiber counter electrode. A larger capacitance (C) was also found in other nanostructured counter electrodes with high porosity (Murakami et al. 2006; Jiang et al. 2010). Unfortunately, the fitted series resistance (R_s) of carbon nanofiber counter electrode was 5.12 Ωcm^2, more than twice of that of 2 Ωcm^2 for Pt electrode. This can be attributed to the higher thickness (~24 μm) of carbon nanofiber counter electrode. It was previously reported that thicker films increase R_s in carbon nanoparticle counter electrodes (Murakami et al. 2006).

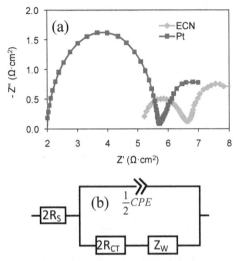

Fig. 6. (a) Nyquist plots of symmetrical carbon nanofiber-carbon nanofiber or Pt-Pt electrode cell; (b) equivalent circuit that was used to fitted the EIS results. Rs is series resistance at the counter electrode, R_{ct} charge transfer resistance, Z_w Nernst diffusion impedance and CPE constant phase element. Reprinted with permission from {Joshi et al. 2010}. Copyright {2010} American Chemical Society.

Counter Electrode	R_s (Ωcm^2)	R_{ct} (Ωcm^2)	C (Fcm^{-2})	β
ECN	5.12	0.70	5.6×10^5	0.82
Pt	2.00	1.89	2.0×10^5	0.95

Table 2. Fitted results extracted from Nyquist plots of the respective symmetrical cells using carbon nanofiber or Pt as electrode. Reprinted with permission from {Joshi et al. 2010}. Copyright {2010} American Chemical Society.

3.3 DSSC performance using carbon nanofiber counter electrode

The TiO_2 photoanode contained a blocking layer, a $TiCl_4$-treated nanocrystalline TiO_2 layer (Solaronix Ti-Nanoxide HT/SP) and a light scattering layer (Dyesol WER4-0). After sintering, the photoanode was soaked in a dye solution made of 0.5 mM Ruthenizer 535-bisTBA dye (Solaronix N-719) in acetonitrile/valeronitrile (1:1). The photoanode was then assembled with carbon nanofiber counter electrode using a thermoplastic sealant. The I^-/I_3^- electrolyte was finally injected into the cells. The reference DSSC devices with sputtered Pt layer (40 nm) as counter electrode were also fabricated for comparison in the same method.

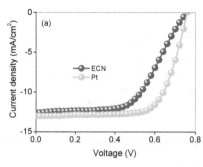 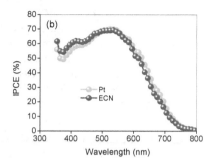

Fig. 7. (a) The J-V curves and (b) IPCE spectral action responses of carbon nanofiber (blue) and Pt (green) counter electrode DSSCs. Reprinted with permission from {Joshi et al. 2010}. Copyright {2010} American Chemical Society.

The J-V curves of carbon nanofiber and Pt DSSCs are shown in Figure 7a, tested under AM 1.5 solar simulator illumination at 100 mWcm^{-2}. Table 3 summaries the device parameters from these two different types of cells. The carbon nanofiber counter electrode DSSCs achieved a J_{sc} of 12.6 mAcm^{-2}, a V_{oc} of 0.76 V, a FF of 0.57, and efficiency (η) of 5.5 %. The corresponding parameters (J_{sc}, V_{oc}, FF, and η) of Pt counter electrode DSSCs were 13.02 mAcm^{-2}, 0.75 V, 0.71, and 6.97 %, respectively. The V_{oc}s of carbon nanofiber and Pt counter electrode DSSCs were very close. The reverse saturation current (J_0) of carbon nanofiber counter electrode DSSCs was 4.47×10^{-9} mAcm^{-2}, comparable to that of 3.58×10^{-9} mAcm^{-2} for Pt counter electrode devices. J_0 is usually regarded as a measure of recombination in solar cells. The comparable value to J_0 suggested that carbon nanofiber counter electrode DSSCs had a comparable recombination as that of Pt counter electrode devices. It was previously reported that charge recombination at FTO/TiO$_2$ and TiO$_2$/electrolyte interfaces in DSSCs led to a V_{oc} loss (Huang et al. 1997; Gratzel 2000; Xia et al. 2007). The comparable Voc in carbon nanofiber and Pt counter electrode DSSCs further conformed that the former did not affect the charge recombination in the DSSCs. However, the J_{sc} is lower in carbon nanofiber counter electrode DSSCs than that of Pt counter electrode DSSCs. Figure 7b shows incident photon-to-current conversion efficiency (IPCE) spectral action responses of the two devices. It was found that IPCE of carbon nanofiber cells was slightly smaller than that of Pt devices in the 550–750 nm spectral range, consistent with the relatively lower J_{sc}. This was probably caused by that the Pt counter electrode can reflect unabsorbed light back to TiO$_2$ photoanode for re-absorption by the dye (Fang et al. 2004; Lee et al. 2009; Wang et al. 2009). However, carbon nanofiber counter electrode cannot reflect such unabsorbed light. However, the reduction of J_{sc} was insignificant and the real reason for lower η in carbon nanofiber cells was the lower FF.

Counter Electrode	J_{sc} (mAcm^{-2})	V_{oc} (V)	FF	η (%)	J_0 (mAcm^{-2})	R_{Stot} (Ωcm^2)
Carbon Nanofiber	12.60	0.76	0.57	5.5	4.47×10^{-9}	15.5
Pt	13.02	0.75	0.71	6.97	3.58×10^{-9}	4.8

Table 3. The comparison of device parameters of carbon nanofiber and Pt counter electrode (R_{Stot}: total series resistance, and J_0: reverse saturation current). Reprinted with permission from {Joshi et al. 2010}. Copyright {2010} American Chemical Society.

The reduced FF may be caused by the increase of overall series resistance (R_{Stot}) of the cells. The R_{Stot} of carbon nanofiber counter electrode was 15.5 Ωcm^2, much larger than that (4.8 Ωcm^2) of Pt devices. Two possible reasons can explain the larger R_{Stot} in carbon nanofiber counter electrode DSSCs. First, the thickness of carbon nanofiber counter electrode (~24 μm) was much thicker than that (40 nm) of Pt electrode. Although the larger thickness provided a larger surface area for triiodide reduction with a reduced R_{ct}, it may increase electron transport length before reaching triiodide reduction sites and lead to higher internal series resistance (Murakami et al. 2006).

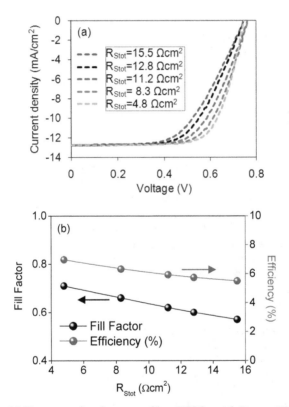

Fig. 8. (a) Simulated J-V curves of carbon nanofiber DSSCs with R_{Stot} at 15.5 Ωcm^2 (blue), 12.8 Ωcm^2 (dark), 11.2 Ωcm^2 (purple), 8.3 Ωcm^2 (red), 4.8 Ωcm^2 (green); (b) Relationship between FF (black), η (blue) and R_{Stot}. Reprinted with permission from {Joshi et al. 2010}. Copyright {2010} American Chemical Society.

The larger internal series resistance was consistent with the higher R_s in carbon nanofiber counter electrode, which was more than twice of that of Pt electrode obtained in symmetrical cells via EIS measurements. Second, carbon nanofiber counter electrode may have a higher Z_W of triiodide ions than Pt electrode because the thicker porous carbon nanofiber film increased the diffusion length of triiodide ions. This can also lead to a larger internal series resistance (Lee et al. 2009). A higher Z_W was also reported previously in other nanostructured counter electrodes including surface-nitrided nickel (Jiang et al. 2010), carbon nanotubes (Lee et al. 2009; Li et al. 2010), and carbon nanoparticles (Murakami et al. 2006). A series of J-V curves at different R_{Stot} were simulated in order to quantitatively study the R_{Stot} effects on carbon nanofiber counter electrode DSSC performance. The dependence of FF and η on R_{Stot} in carbon nanofiber counter electrode is plotted in Figure 8b. When decreasing R_{Stot} from 15.5 to 4.8 Ωcm^2, FF and η can be significantly improved and approach that of Pt DSSCs. A promising approach to decreasing R_{Stot} is to reduce the thickness of carbon nanofiber counter electrode via a thinner and more porous carbon nanofiber sheet.

4. Conclusion

The carbon/TiO_2 composite and carbon nanofiber were used as low cost alternative to Pt counter electrode for DSSCs. Although the efficiencies of carbon/TiO_2 composite and carbon nanofiber DSSC devices were lower than that of Pt devices, some of the parameters including J_{sc}, V_{oc} or FF are comparable. The carbon/TiO_2 composite and carbon nanofiber counter electrodes have shown potential as an efficient electrocatalyst with low charge transfer resistance (R_{ct}) and large surface area for reduction of I_3^- ions.

5. Acknowledgment

This work was mainly supported by NSF EPSCoR/PANS program (EPS-EPSCoR-0903804 with some shared materials from NSF CAREER (ECCS-0950731) and NASA EPSCoR (NNX09AP67A). This work was modified from the previous journal publicaions (Joshi et al. 2009; Joshi et al. 2010). The authors acknowledge Drs Hao Fong and Lifeng Zhang at the South Dakota School of Mines and Technologies to provide carbon nanofiber samples for DSSC applications. The authors are also grateful to Dr. Mahdi F. Baroughi for help in setting up the J-V and IPCE measurement systems in the Department of Electrical Engineering and Computer Science at the South Dakota State University.

6. References

Alibabaei, L., M. K. Wang, et al. (2010). "Application of Cu(II) and Zn(II) Coproporphyrins as Sensitizers for Thin Film Dye Sensitized Solar Cells." *Energy & Environmental Science* 3(7): 956-961.

Burnside, S., S. Winkel, et al. (2000). "Deposition and Characterization of Screen-Printed Porous Multi-layer Thick Film Structures from Semiconducting and Conducting Nanomaterials for Use in Photovoltaic Devices." *Journal of Materials Science - Materials in Electronics* 11(4): 355-362.

Calandra, P., G. Calogero, et al. (2010). "Metal Nanoparticles and Carbon-Based Nanostructures as Advanced Materials for Cathode Application in Dye-Sensitized Solar Cells." *International Journal of Photoenergy* 2010: 109495.

Fan, B. H., X. G. Mei, et al. (2008). "Conducting Polymer/Carbon Nanotube Composite as Counter Electrode of Dye-Sensitized Solar Cells." *Applied Physics Letters* 93(14).

Fang, X. M., T. L. Ma, et al. (2004). "Effect of the Thickness of the Pt Film Coated on a Counter Electrode on the Performance of a Dye-Sensitized Solar Cell." *Journal of Electroanalytical Chemistry* 570(2): 257-263.

Gajjela, S. R., K. Ananthanarayanan, et al. (2010). "Synthesis of Mesoporous Titanium Dioxide by Soft Template Based Approach: Characterization and Application in Dye-Sensitized Solar Cells." *Energy & Environmental Science* 3(6): 838-845.

Gratzel, M. (2000). "Perspectives for Dye-Sensitized Nanocrystalline Solar Cells." *Progress in Photovoltaics: Research and Applications* 8(1): 171-185.

Gratzel, M. (2003). "Dye-sensitized solar cells." *Journal of Photochemistry and Photobiology C-Photochemistry Reviews* 4(2): 145-153.

Han, H. W., U. Bach, et al. (2009). "A Design for Monolithic All-Solid-State Dye-Sensitized Solar Cells with a Platinized Carbon Counterelectrode." *Applied Physics Letters* 94(10).

Hauch, A. and A. Georg (2001). "Diffusion in the electrolyte and charge-transfer reaction at the platinum electrode in dye-sensitized solar cells." *Electrochimica Acta* 46(22): 3457-3466.

Hinsch, A., S. Behrens, et al. (2008). "Material Development for Dye Solar Modules: Results from an Integrated Approach." *Progress in Photovoltaics: Research and Applications* 16(6): 489-501.

Huang, S. Y., G. Schlichthorl, et al. (1997). "Charge Recombination in Dye-Sensitized Nanocrystalline TiO2 Solar Cells." *Journal of Physical Chemistry B* 101(14): 2576-2582.

Huang, Z., X. Liu, et al. (2007). "Application of Carbon Materials as Counter Electrodes of Dye-Sensitized Solar Cells." *Electrochemistry Communications* 9(4): 596-598.

Imoto, K., M. Suzuki, et al. (2003). "Activated carbon counter electrode for dye-sensitized solar cell." *Electrochemistry* 71(11): 944-946.

Imoto, K., K. Takahashi, et al. (2003). "High-performance carbon counter electrode for dye-sensitized solar cells." *Solar Energy Materials and Solar Cells* 79(4): 459-469.

Jiang, Q. W., G. R. Li, et al. (2010). "Surface-Nitrided Nickel with Bifunctional Structure As Low-Cost Counter Electrode for Dye-Sensitized Solar Cells." *The Journal of Physical Chemistry C* 114(31): 13397-13401.

Joshi, P., Y. Xie, et al. (2009). "Dye-Sensitized Solar Cells based on Low Cost Nanoscale Carbon/TiO2 Composite Counter Electrode." *Energy & Environmental Science* 2(4): 426-429.

Joshi, P., L. Zhang, et al. (2010). "Electrospun Carbon Nanofibers as Low-Cost Counter Electrode for Dye-Sensitized Solar Cells." *ACS Applied Materials & Interfaces* 2(12): 3572-3577.

Kay, A. and M. Gratzel (1996). "Low Cost Photovoltaic Modules Based on Dye Sensitized Nanocrystalline Titanium Dioxide and Carbon Powder." *Solar Energy Materials and Solar Cells* 44(1): 99-117.

Lee, W. J., E. Ramasamy, et al. (2009). "Efficient Dye-Sensitized Solar Cells with Catalytic Multiwall Carbon Nanotube Counter Electrodes." *ACS Applied Materials & Interfaces* 1(6): 1145-1149.

Li, G.-R., F. Wang, et al. (2010). "Carbon Nanotubes with Titanium Nitride as a Low-Cost Counter-Electrode Material for Dye-Sensitized Solar Cells13." *Angewandte Chemie International Edition* 49(21): 3653-3656.

Matsubara, T., R. Sakaguchi, et al. (2005). "Measurement and Analysis of the Series Resistance in a Dye Sensitized Solar Cells " *Electrochemistry* 73(1): 60 - 66.

Mei, X. and J. Ouyang (2009). "Gels of carbon nanotubes and a nonionic surfactant prepared by mechanical grinding." *Carbon* 48: 293-299.

Murakami, T. N., S. Ito, et al. (2006). "Highly Efficient Dye-Sensitized Solar Cells based on Carbon Black Counter Electrodes." *Journal of the Electrochemical Society* 153(12): A2255-A2261.

Oregan, B. and M. Gratzel (1991). "A Low-Cost, High-Efficiency Solar-Cell Based on Dye-Sensitized Colloidal Tio2 Films." *Nature* 353(6346): 737-740.

Papageorgiou, N., W. F. Maier, et al. (1997). "An Iodine/Triiodide Reduction Electrocatalyst for Aqueous and Organic Media." *Journal of The Electrochemical Society* 144(3): 876-884.

Ramasamy, E., W. J. Lee, et al. (2007). "Nanocarbon counterelectrode for dye sensitized solar cells." *Applied Physics Letters* 90(17): 173103.

Skupien, K., P. Putyra, et al. (2009). "Catalytic Materials Manufactured by the Polyol Process for Monolithic Dye-sensitized Solar Cells." *Progress in Photovoltaics: Research and Applications* 17(1): 67-73.

Sun, H., Y. Luo, et al. "In Situ Preparation of a Flexible Polyaniline/Carbon Composite Counter Electrode and Its Application in Dye-Sensitized Solar Cells." *The Journal of Physical Chemistry C* 114(26): 11673-11679.

Sun, K., B. H. Fan, et al. (2010). "Nanostructured Platinum Films Deposited by Polyol Reduction of a Platinum Precursor and Their Application as Counter Electrode of Dye-Sensitized Solar Cells." *Journal of Physical Chemistry C* 114(9): 4237-4244.

Suzuki, K., M. Yamaguchi, et al. (2003). "Application of carbon nanotubes to counter electrodes of dye-sensitized solar cells." *Chemistry Letters* 32(1): 28-29.

Wang, M., A. M. Anghel, et al. (2009). "CoS Supersedes Pt as Efficient Electrocatalyst for Triiodide Reduction in Dye-Sensitized Solar Cells." *Journal of the American Chemical Society* 131(44): 15976-15977.

Xia, J., N. Masaki, et al. (2007). "Sputtered Nb2O5 as a Novel Blocking Layer at Conducting Glass/TiO2 Interfaces in Dye-Sensitized Ionic Liquid Solar Cells." *The Journal of Physical Chemistry C* 111(22): 8092-8097.

Mei, X., S. J. Cho, et al. (2010). "High-performance dye-sensitized solar cells with gel-coated binder-free carbon nanotube films as counter electrode." *Nanotechnology* 21(39): 395202.

Xie, Y., P. Joshi, et al. (2010). "Electrolyte Effects on Electron Transport and Recombination at ZnO Nanorods for Dye-Sensitized Solar Cells." *The Journal of Physical Chemistry C* 114(41): 17880-17888.

Xie, Y., P. Joshi, et al. (2009). "Structural effects of core-modified porphyrins in dye-sensitized solar cells." *Journal of Porphyrins and Phthalocyanines* 13(8-9): 903-909.

Yum, J. H., E. Baranoff, et al. (2010). "Phosphorescent Energy Relay Dye for Improved Light Harvesting Response in Liquid Dye-Sensitized Solar Cells." *Energy & Environmental Science* 3(4): 434-437.

9

Fabrication of ZnO Based
Dye Sensitized Solar Cells

A.P. Uthirakumar
Nanoscience Centre for Optoelectronics and Energy Devices,
Sona College of Technology, Salem, Tamilnadu,
India

1. Introduction

Why solar power is considered as one of the ultimate future energy resources? To answer this, drastic depletion of fossil fuels and the challenges ahead on needs for the specific requirements are the major causes for the need alternative power. At the beginning of February, 2007, the Intergovernmental Panel on Climate Change (*IPCC*) presented a report concluding that global concentrations of carbon dioxide, methane and nitrous oxide have increased markedly as a result of human activities since 1750. The report states that the increase in carbon dioxide, the most important greenhouse gas, is primarily due to fossil fuel use. The report further indicates that the increased concentrations of carbon dioxide, methane, and nitrous oxide have increased the average global temperature, a phenomenon known as "global warming". Eventually, if the temperature continues to increase, this will influence our everyday lives, since it changes the conditions of, for example, agriculture and fishing. In order to preserve the surplus energy coming out from the Sun, an alternative technique is necessary for our future energy needs. The potential of using the sun as a primary energy source is enormous. For example, sunlight strikes the Earth in one hour (4.3×1020 J) is sufficient to satisfy the more than the globel energy consumed on the planet in a year (4.1×1020 J). In other words, it has been calculated that covering 0.1% of the earth's surface area with solar cells of 10% efficiency, corresponding to 1% of desert areas or 20% of the area of buildings and roads, would provide for global electricity consumption. As for as to convert the solar power into the basic electricity, there may be the new technology to be implimented to harvest the solar energy in an effective manner. In this regard, one way to consume solar power, the photovoltaic cell will be suitable one to convert sunlight into electrical energy. The challenge in converting sunlight into electricity via photovoltaic solar cells is dramatically reducing the cost/watt of delivered solar electricity, by approximately a factor of 5-10 times to compete with fossil and nuclear electricity and by a factor of 25-50 to compete with primary fossil energy.

Recently, many of research groups are actively involving to harvest maximum conversion of solar power into electricity. Hence, varieties of new materials that are capable to absorb solar spectrum are successfully prepared in different methods. These new materials should satisfy the following important points to be act as the effective light harvesting materials. It should be efficiently absorb sunlight, should cover the full spectrum of wavelengths in solar radiation, and new approaches based on nanostructured architectures can revolutionize the technology used to produce energy from the solar radiation. The technological development in novel approaches exploiting thin films, organic semiconductors, dye sensitization, and quantum dots offer fascinating new opportunities for cheaper, more efficient, longer-lasting systems. The conversion from solar energy to electricity is fulfilled by solar-cell devices based on the photovoltaic effect. Many photovoltaic devices have already been developed over the past five decades (Liu et al., 2008). However, wide-spread use is still limited by two significant challenges, namely conversion efficiency and cost (Bagnall et al., 2008). One of the traditional photovoltaic devices is the single-crystalline silicon solar cell, which was invented more than 50 years ago and currently makes up 94% of the market (Chapin et al., 1954). In addition to this other compound semiconductors, such as gallium arsenide (GaAs), cadmium telluride (CdTe), and copper indium gallium selenide, receive much attention because they present direct energy gaps, can be doped to either p-type or n-type, have band gaps matching the solar spectrum, and have high optical absorbance (Afzaal & O'Brien, 2006). These devices have demonstrated single-junction conversion efficiencies of 16–32% (Birkmire, 2001). Although those photovoltaic devices built on silicon or compound semiconductors have been achieving high efficiency for practical use, they still require major breakthroughs to meet the long-term goal of very-low cost.

In case of lowering the cost of production, dye-sensitized solar cells (DSSCs) based on oxide semiconductors and organic dyes or metallorganic-complex dyes have recently emerged as promising approach to efficient solar-energy conversion. The DSSCs are a photoelectrochemical system, which incorporate a porous-structured oxide film with adsorbed dye molecules as the photosensitized anode. A typical DSSC system composed of a mesoporous titanium dioxide (TiO_2) film on a transparent conductor. Dye molecules are absorbed on the entire porous TiO_2 that is perfused with an electrolyte containing iodide and tri-iodide. A layer of additional electrolyte separates the porous TiO_2 from a counter electrode. When a photon is absorbed by a dye, the excited dye transfers an electron to the TiO_2 (termed injection). The then oxidized dye can be reduced by iodide (regeneration) or can recapture an electron from the TiO_2. The electron in the TiO_2 can diffuse to a collection electrode (transport) or can be captured by a triiodide molecule in the electrolyte. Electrons that reach the collection electrode flow through the external circuit and reduce tri-iodide to iodide at the counter electrode.

Compared with the conventional single-crystal silicon-based or compound-semiconductor thin-film solar cells, DSSCs are thought to be advantageous as a photovoltaic device possessing both practicable high efficiency and cost effectiveness. To date, the most successful DSSC was obtained on TiO_2 nanocrystalline film combined with a ruthenium-polypyridine complex dye (Grätzel et al., 2000, 2001, 2003 & 2007). Following this idea, a certified overall conversion efficiency of 10.4% was achieved on a TiO_2–RuL'(NCS)3 ("black dye") system, in which the spectral response of the complex dye was extended into the near-infrared region so as to absorb far more of the incident light. The porous nature of nanocrystalline TiO_2 films drives their use in DSSCs due to the large surface area available for dye-molecule adsorption. Meanwhile, the suitable relative energy levels at the semiconductor–sensitizer interface (i.e., the position of the conduction-band edge of TiO2 being lower than the excited-state energy level of the dye) allow for the effective injection of electrons from the dye molecules to the semiconductor (Nelson & Chandler, 2004).

2. The dye-sensitized solar cell

A promising alternative energy source of DSSC will be expected to increase the significant contribution to overall energy production over the coming years. This is mainly due to the offering a low-cost fabrication and attractive features such as transparency, flexibility, etc.

that might facilitate the market entry. Among all of them, DSSCs are devices that have shown to reach moderate efficiencies, thus being feasible competitors to conventional cells. DSSC combine the optical absorption and charge-separation processes by the association of a sensitizer as light-absorbing material with a wide band-gap semiconductor (usually titanium dioxide). A schematic representation of energy flow in DSSC is illustrated in Figure 1. As early as the 1970s, it was found that TiO_2 from photoelectrochemical cells could split water with a small bias voltage when exposed to light (Fujishima & Honda, 1972). However, due to the large band-gap for TiO_2, which makes it transparent for visible light, the conversion efficiency was low when using the sun as illumination source. This pioneering research involved an absorption range extension of the system into the visible region, as well as the verification of the operating mechanism by injection of electrons from photoexcited dye molecules into the conduction band of the n-type semiconductor. Since only a monolayer of adsorbed dye molecules was photoactive, light absorption was low and limited when flat surfaces of the semiconductor electrode were employed. This inconvenience was solved by introducing polycrystalline TiO_2 (anatase) films with a surface roughness factor of several hundreds Desilvestro et al., 1985). The amount of adsorbed dye was increased even further by using mesoporous electrodes, providing a huge active surface area thereby, and cells combining such electrodes and a redox electrolyte based on iodide/triiodide couple yielded 7% conversion efficiencies in 1991 (Nazeerudin et al., 1993). The next highest energy conversion efficiency is over 11% (Chiba et al., 2006), and further increase of the efficiency is possible by designing proper electrodes and sensitization dyes. Michael Grätzel is a professor at the École Polytechnique Fédérale de Lausanne where he directs the Laboratory of Photonics and Interfaces. He pioneered research on energy and electron transfer reactions in mesoscopic-materials and their optoelectronic applications and discovered a new type of solar cell based on dye sensitized mesoscopic oxide particles and pioneered the use of nanomaterials in lithium ion batteries. Today, record efficiencies above 13% have been presented at an illumination intensity of 1000 Wm-2, and the device displays promising stability data (Gra¨tzel et al., 2000, 2001, 2003 & 2007).

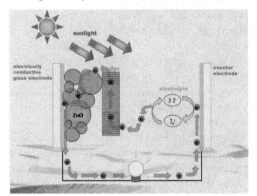

Fig. 1. Schematic diagram of the energy flow in the dye-sensitized solar cell.

The possible mechanism of electron hole interaction during DSSC operation is illustrated in Figure 2. After the successful excitation of dye molecules with the sun light, the ground state electrons at the highest occupied molecular orbital will be photoexcited and residing at unoccupied molecular orbital. Then, the photoexcited electrons are passing through the nanomaterials to reach electrodes for completion. The detailed mechanism for the electron transports within DSSC device is described as shown in Figure 2. Here, photoexcited electrons are injected from the dye to the conduction band (denoted as "c.b.") of the nanocrystallite (1), the dye is regeneratedby electron transfer from a redox couple in the electrolyte (3), and a recombination may take place between the injected electrons and the dye cation (2) or redox couple (4). The latter (4) is normally believed to be the predominant loss mechanism. Electron trapping in the nanocrystallites (5) is also a mechanism that causes energy loss.

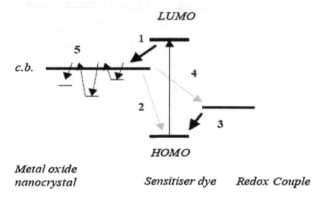

Fig. 2. A schematic illustrates the electron-hole interactions after the excitation of dye molecules with the sun light.

In the recent years, the DSSCs offer to achieve moderate conversion efficiency to the maximum of ~14%. In this case, a wide band gap, mesoporous nanocrystalline TiO_2 was used as photoanodic material, on which dye molecules are adsorbed for the photo harvesting. Major achievers in the DSSC developments are mainly from the recent technology development and optimization. Now in the place of TiO_2, an another alternative, a direct wide band gap metal oxide material is zinc oxide (ZnO), a wurtzite type semiconductor with an energy gap of 3.37 eV at room temperature, as summerziaed in Table 1. Due to its large bandgap, ZnO is an excellent semiconductor material like other wide bandgap materials of GaN and SiC. The crystal strcture, energy band gap, electron mobility and electron diffusion coefficient of both ZnO and TiO_2 nanomaterials were summerized in Table 1, for comparison. The band gap energy of ZnO nanomaterials is almost same as that of TiO_2 and the electron mobility and electron diffusion coefficient of ZnO showed much highier values than the TiO_2, which facilites the important for the DSSC application.

To understand this issue, DSSC technology based on ZnO has been explored extensively. ZnO is a wide-band-gap semiconductor that possesses an energy-band structure and physical properties similar to those of TiO_2 (Table 1), but has higher electronic mobility that would be favorable for electron transport, with reduced recombination loss when used in DSSCs. Hence, many studies have already been started and reported on the useage of ZnO nanomaterial for application in DSSCs. Although the conversion efficiencies of 0.4–5.8% obtained for ZnO are much lower than that of 11% for TiO_2, ZnO is still thought of as a distinguished alternative to TiO_2 due to its ease of crystallization and anisotropic growth. These properties allow ZnO to be produced in a wide variety of nanostructures, thus presenting unique properties for electronics, optics, or photocatalysis (Tornow et al., 2007 & 2008 and Djurisic et al 2006). Parallely, numerous works have been published to explore in-depth analysis of their shape controllable synthesis that includes needles, rods, tubes, towers, hollow prisms, tetra-legs, flowers, stars, helices, belts and springs via simple method. For example, the major reason for useage of nanostructures with a large specific surface area help in many particular behaviors in electron transport or light propagation in view of the surface effect, quantum-confinement effect or photon localization (Bittkau et al., 2007). Those nanostructural forms of ZnO developed during the past several decades mainly include nanoparticles, nanorod, nanotubes, nanobelts, nanosheets and nanotips. The production of these structures can be achieved through sol–gel synthesis, hydrothermal /solvothermal growth, physical or chemical vapor deposition, low-temperature aqueous growth, chemical bath deposition, or electrochemical deposition.

	ZnO	TiO$_2$
Crystal structure	rocksalt, zinc blende, and wurtzite	rutile, anatase, and brookite
Energy band gap [eV]	3.2–3.3	3.0–3.2
Electron mobility [cm^2 Vs^1]	205–300 (bulk ZnO), 1000 (single nanowire)	0.1–4
Refractive index	2.0	2.5
Electron effective mass[me]	0.26	9
Relative dielectric constant	8.5	170
Electron diffusion coefficient [cm^2 s^1]	5.2 (bulk ZnO), 1.7 $_x$ 10^4 (nano-particulate film)	0.5 (bulk TiO$_2$), 10^8–10^4 (nano- particulate film)

Table 1. A comparison of physical properties of ZnO and TiO$_2$.

In particular, recent studies on ZnO-nanostructure-based DSSCs have delivered many new concepts, leading to a better understanding of photoelectrochemically based energy conversion. This, in turn, would speed up the development of DSSCs that are associated with TiO$_2$. Moreover, these ZnO nanomaterials can be synthesized through simple chemical methods with the wide range of structural evolution, fabricating DSSCs with ZnO nano-structured materials will be advisable and reliable in place of TiO$_2$, whose structural controllability is not easy in a conversional chemical synthetic route. In this chapter, recent developments in ZnO nanostructures, particularly for application in DSSCs, are reported. It will show that photoelectrode films with nanostructured ZnO can significantly enhance solar-cell performance by offering a large surface area for dye adsorption, direct transport pathways for photoexcited electrons, and efficient scattering centers for enhanced light-harvesting efficiency. The limitations of ZnO-based DSSCs are also discussed. In the final section, several attempts to expand ZnO concepts to TiO$_2$ are presented to motivate further improvement in the conversion efficiency of DSSCs.

3. Crystal and surface structure of ZnO

At ambient pressure and temperature, ZnO crystallizes in the wurtzite (B4 type) structure, as shown in Figure 3 (Jagadish et al., 2006). This is a hexagonal lattice, belonging to the space group P$_{63}$mc with lattice parameters a = 0.3296 and c = 0.52065 nm. Usually, we can treat it as a number of two type planes, i.e. tetrahedrally coordinated O^{2-} and Zn^{2+} ions, and stacked alternately along the c-axis. Or in another way, it also can be characterized by two interconnecting sublattices of Zn^{2+} and O^{2-}, such that each Zn ion is surrounded by tetrahedra of O ions, and vice-versa. No doubt, this kind of tetrahedral coordination in ZnO will form a noncentral symmetric structure with polar symmetry along the hexagonal axis, which not only directly induces the characteristic piezoelectricity and spontaneous polarization, but also plays a key factor in crystal growth, etching and defect generation of ZnO.

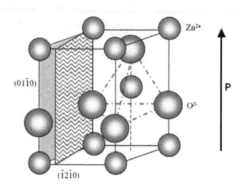

Fig. 3. The wurtzite structure model of ZnO. The tetrahedral coordination of Zn–O is shown.

The polar surface is another important characteristic of ZnO structure. As well as we known, wurtzite ZnO crystallizes do not have a center of inversion. If the ZnO crystals such as nanorods and nanotubes grow along the c axis, two different polar surfaces will be formed on the opposite sides of the crystal due to the suddenly termination of the structure, i.e. the terminated Zn-(0001) surface with Zn cation outermost and the terminated O-(000) surface with O anion outermost. Naturally, these positively charged Zn-(0001) and negatively charged O-(000) surfaces are the most common polar surfaces in ZnO, which subsequently results in a normal dipole-moment and spontaneous polarization along the c-axis as well as a divergence in surface energy. Generally, the polar surfaces have facets or exhibit massive surface reconstructions in order to maintain a stable structure. However, ZnO-± (0001) are exceptions: they are atomically flat, stable and without reconstruction (Dulub et al., 2002 & Wander et al. 2001). Efforts to understand the superior stability of the ZnO ± (0001) polar surfaces are at the forefront of research in today's surface physics (Staemmler et al, 2003 & Singh et al., 2007).

4. Basic physical parameters for ZnO

The basic physical parameters of ZnO were compiled and placed as in Table 2 (Pearton et al., 2005 & Florescu et al., 2002). The physical parameters include energy band gap, excitons binding energy, electron effective mass, hole effective mass, hole hall mobility and lattice paramerter for easy understanding. It should also be noted that there still exists uncertainty in some of these values. For example, there have few reports of p-type ZnO and therefore the hole mobility and effective mass are still in debate. Similarly, the values for thermal conductivity show some spread in values and this may be a result of the influence of defects such as dislocations (Hosokawa et al., 2007), as was the case for GaN. The values for carrier mobility will undoubtedly increase as more control is gained over compensation and defects in the material.

Physical parameters	Values
Stable phase at 300 K	Wurtzite
Energy band gap	3.37 eV, direct
Density	5.606 g/cm^3
Refractive index	2.008, 2.029
Melting point	1975°C
Thermal conductivity	0.6, 1–1.2
Exciton binding energy	60 meV
Electron effective mass	0.24
Electron Hall mobility at 300 K for low n-type conductivity	200 cm^2/V s
Linear expansion coefficient(/^{o}C)	a_0: 6.5 ×10^{-6} and c_0: 3.0 × 10^{-6}
Static dielectric constant	8.656
Hole effective mass	0.59
Hole Hall mobility at 300 K for low p-type conductivity	5–50 cm^2/V s
Intrinsic carrier concentration	<106 cm^{-3} (max n-type doping >1020 cm^{-3} electrons; max p-type doping <1017 cm^{-3} holes)
Lattice parameters at 300 K	
a_0	0.32495 nm
c_0	0.52069 nm
a_0/c_0	1.602 (ideal hexagonal structure shows 1.633)
U	0.345

Table 2. Physical parameters of ZnO.

5. Synthesis of ZnO nanomaterials for DSSC

The major objective of selecting a photoanodic nanomaterials film for DSSC, it should offer large internal surface area whereby to adsorb sufficient dye molecules for the effect capture of incident photons from the solar power. This objective will be solved by the formation of a porous interconnected network in which the specific surface area may be increased by more than 1000 times when compared with bulk materials (Lee et al., 2003). In this respect, ZnO is a key technological material and it has a wide band-gap compound semiconductor that is suitable for optoelectronic applications. In addition to this, the abundant forms of ZnO nanostructures provide a great deal of opportunities to obtain high surface-area-to-volume ratios, which helps to contribute the successful dye adsorption leading to a better light harvesting, in DSSCs. Therefore, many research groups are actively intriguing to dealings with this task of preparing various structurally different types of ZnO nanostructures to fabricate the DSSC for future energy crisis. Hence, variety of different synthetic methods, such as vapor-phase transport (Zhao et al., 2007, Huang et al., (2001) & Sun et al., (2004), pulsed laser deposition (Wu et al., 2002), chemical vapor deposition (Park et al., 2002, Yu et al., 2005) and electrochemical deposition (Zeng et al., 2006).

Structure	Ru based dyes	Efficiency
Nanoparticles	N719	0.44%, 2.1% (0.06 sun), 2.22%
	N719	5% (0.1 sun)
	N3	0.4%, 0.75%, 2% (0.56 sun), 3.4%
Nanorods	N719	0.73%
	N719	0.22%
	N719	1.69%
Nanotips	N719	0.55%, 0.77%
Nanotubes	N719	1.6%, 2.3%
Nanobelts	N719	2.6%
Nanosheets	N719	2.61%, 3.3%
	N3	1.55%
Nanotetrapods	N719	1.20%, 3.27%
Nanoflowers	N719	1.9%
Nanoporous films	N3	5.08% (0.53 sun)
	N719	3.9%, 4.1%
	N719	0.23%
Nanowires	N3	0.73%, 2.1%, 2.4%, 4.7%
	N719	0.3%, 0.6%, 0.9%,1.5%, 1.54%
Aggregates	N3	3.51%, 4.4%, 5.4%

Table 3. Summary of DSSCs based on ZnO nanostructures.

ZnO nanostructured materials with diverse range of structureally distinct morphologies were synthesized from different methods as listed in Table 3. The detailed behind the morphologically distinct ZnO nanomaterials utilization in the DSSC application with the help of Ru dye complex and their impact of solar power geneacration also displayed in Table 3. The followings are the few examples of diverse group of ZnO growth morphologies, such as nanoparticles (Keis et al., 2000, Suliman et al., 2007 & Gonzalez-Valls et al., 2010), nanorod (Lai et al., 2010, Hsu et al., 2008 & Charoensirithavorn et al., 2006), nanotips (Martinson et al., 2007), nanotubes (Lin et al., 2008), nanobelts (Kakiuchi et al., 2008), nanosheets (Chen et al., 2006) , nanotetrapods (Jiang et al, 2007), nanoflowers (Chen et al., 2006), nanoporous films (Hosono et al., 2005, Kakiuchi et al., 2006 & Guo et al., 2005), nanowires (Guo et al., 2005, Rao et al., 2008, Wu et al., 2007 & Law et al., 2005) and aggregates (Chou et al., 2007 & Zhang et al., 2008). These ZnO nanostructures are easily prepared even on cheap substrates such as glass and utilized for the DSSC application as photoanodic materials. Hence, they have a promising potential in the nanotechnology future. The specific impact of individually distinct ZnO nanomaterials will be discussed in the subsequent section in details.

5.1 Useage of ZnO nanoparticles as photoanodic material

The first and foremost interest on ZnO structutural morphology is of spherical shaped nanoparticles (NPs) that suit in both synthetic methodoly as well as process simplicity. In particular, ZnO NPs can easily be prepared in simple methods by proper judification of their reaction conditions and parameters. Uthirakumar et al. reported simple solution method for the preparation of verity of ZnO nanostructured materials out of which is ZnO NP, one of the significant nanomaterials. The diverse morphology of ZnO nanostructures synthesized from solution method is displayed in Figure 4. They continued to utilize these nanoparticles for

DSSC device fabrication (Uthirakumar et al., 2006, 2007, 2008 & 2009). ZnO NPs with N3 dye sensitizer produced the higher solar power conversion efficiency ranges from 0.44 to 3.4% (Keis et al., 2000, Uthirakumar et al., 2009). However, further improvement of maximum of 5% conversion efficiency with N719 dye. Hosono et al systematically studied the DSSC performance of nanoporous structured ZnO films fabricated by the CBD technique (Hosono et al., 2004, 2005 & 2008). They achieved an overall conversion efficiency of 3.9% when as-prepared 10-mm-thick ZnO films were sensitized by N719 dye with an immersion time of 2 h (Hosono et al., 2005). Further improvement to 4.27% in the conversion efficiency was reported recently by Hosono et al. when the dye of N719 was replaced with a metal-free organic dye named D149 and the immersion time was reduced to 1h (Hosono et al., 2008). The enhancement in solar-cell performance was attributed to the use of D149 dye and a nanoporous structure that contained perpendicular pores. This allowed for a rapid adsorption of the dye with a shorter immersion time and thus prevented the formation of a $Zn^2þ$/dye complex. This complex is believed to be inactive and may hinder electron injection from the dye molecules to the semiconductor [66]. In another study, a high photovoltaic efficiency of up to 4.1% was also obtained for nanoporous ZnO films produced by the CBD (Kakiuchi et al., 2006). However, the excellence of the solar-cell performance was ascribed to the remarkably improved stability of as-fabricated ZnO films in acidic dye.

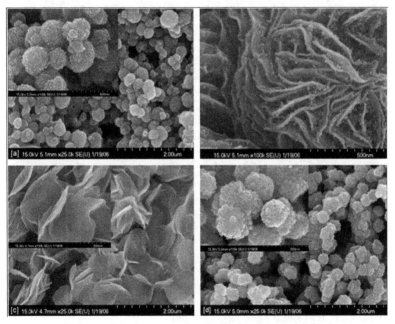

Fig. 4. Diverse morphology of ZnO nanomaterials from solution method.

5.2 Useage of ZnO nanorods as photoanodic material

Controllable length of ZnO nanorods can be grown in solution. The ZnO nanorods are formed at a relatively high temperature (~90 °C), where the reaction solution is enriched with colloidal $Zn(OH)_2$ and therefore allows a fast growth of ZnO nanocrystals along the [001] orientation to form nanorods. ZnO nanorods were grown on the seeded substrates in a sealed chemical bath containing 10 mM each of zinc nitrate ($Zn[NO_3]_2 \cdot 6H2O$) and hexamine ($[CH2]_6N_4$) for 15 h at 90 °C. Photoanodic ZnO nanorod electrodes can be made with vertically-aligned ZnO nanorods and analyzed the usage of DSSC. The highest solar cell efficiency obtained was 0.69% after UV light irradiation (at 72 °C, 0.63 V, 2.85 mA cm^{-2}, 0.39 FF) (Gonzalez-Valls et al., 2010). Typical nanorod-based DSSCs are fabricated by growing nanorods on top of a transparent conducting oxide, as shown in Figure 5. The heterogeneous interface between the nanorod and TCO forms a source for carrier scattering. The new DSSCs yield a power conversion efficiency of 0.73% under 85 mW/cm2 of simulated solar illumination (Lai et al., 2010). Hydrothermally grown and vapor deposited nanorods also exhibit different dependence of photovoltaic performance on the annealing conditions of the rods, indicating significant effect of the native defects on the achievable photocurrent and power conversion efficiency. Efficiency of 0.22% is obtained for both as grown hydrothermally grown nanorods and vapor deposited nanorods annealed in oxygen at 200°C (Hsu et al., 2008). P. Charoensirithavorn et al., proposed a new possibility in designing a cell structure produced an open circuit voltage (Voc) of 0.64 mV, a short circuit current density (Jsc) of 5.37 mA/cm2, a fill factor (FF) of 0.49, and conversion efficiency (η) of 1.69 %, primarily limited by the surface area of the nanorod array (sirithavorn et al., 2006).

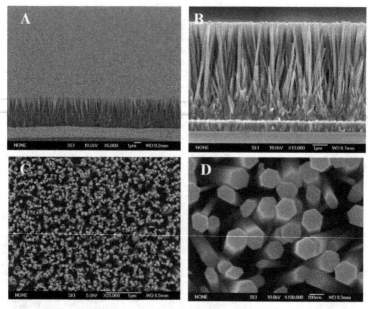

Fig. 4. SEM images of ZnO nanorods grown on FTO substrate (A) tilt, (B) side and (C) top and (D) bottom view at low and high magnification.

5.3 Usage of ZnO nanotubes as photoanodic material

Among one-dimensional ZnO nanostructures, the tubular structures of ZnO become particularly important in DSSC are required their high porosity and large surface area to fulfill the demand for high efficiency and activity. A subsequent decrease in the temperature yields a supersaturated reaction solution, resulting in an increase in the concentration of OH$^-$ ions as well as the pH value of the solution. Colloidal $Zn(OH)_2$ in the supersaturated solution tends to precipitate continually. However, because of a slow diffusion process in view of the low temperature and low concentration of the colloidal Zn(OH)$_2$, the growth of nanorods is limited but may still occur at the edge of the nanorods due to the attraction of accumulated positive charges to those negative species in the solution, ultimately leading to the formation of ZnO nanotubes, as clearly represented in Figure 5(a). The role of changing the pH value observed in the growth of ZnO crystals is shown also to have a relationship to the change of the surface energy. In the course of growing ZnO nanorods, changing the growth temperature, from a high (90 °C) to a low temperature (60 °C), leads to some change in the pH value. At the low pH value, the polar face has such a high surface energy that it permits the growth of nanorods. However, the grain growth can be inhibited by a high pH value at a low growth temperature. The competition between the change of surface energy due to pH value and growth rate dictated by the temperature can be assumed to lead to the ZnO tube structure, as shown in Figure 5(b). This investigation provides more options and flexibility in controlling methods to obtain various morphologies of ZnO crystals in terms of the change of growth temperature and pH value. Other synthetic methods for the preparation of nanotubes are realized by electrochemical method, low temperature solution method, vapor phase growth and the simple chemical etching process to convert the nanorods into nanotubes. The chemical etching process was carried out by suspending the nanorods sample upside down in 100 ml aqueous solution of potassium chloride (KCl) with 5M concentration for 10 h at 95 °C.

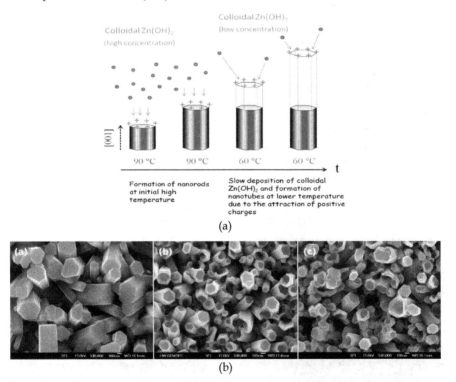

(a)

(b)

Fig. 5. A schematic representation and B) SEM images on evolution of ZnO nanorods to tubes while the solution was kept at 90 °C for 3 h and then cooled down to (a) 80 °C (20 h), (b) 60 °C (20 h) and (c) 50 °C (20 h).

High-density vertically aligned ZnO nanotube arrays were fabricated on FTO substrates by a simple and facile chemical etching process from electrodeposited ZnO nanorods. The nanotube formation was rationalized in terms of selective dissolution of the (001) polar face. The morphology of the nanotubes can be readily controlled by electrodeposition parameters for the nanorod precursor. By employing the 5.1 μm-length nanotubes as the photoanode for a DSSC, a full-sun conversion efficiency of 1.18% was achieved (Han et al., 2010). Alex et al introduce high surface area ZnO nanotube photoanodes templated by anodic aluminum oxide for use in dye-sensitized solar cells (DSSCs). Compared to similar ZnO-based devices, ZnO nanotube cells show exceptional photovoltage and fill factors, in addition to power efficiencies up to 1.6%. The novel fabrication technique provides a facile, metal-oxide general route to well-defined DSSC photoanodes (Martinson et al., 2010). Nanotubes differ from nanowires in that they typically have a hollow cavity structure. An array of nanotubes possesses high porosity and may offer a larger surface area than that of nanowires. An overall conversion efficiency of 2.3% has been reported for DSSCs with ZnO nanotube arrays possessing a nanotube diameter of 500 nm and a density of 5.4 x10^6 per square centimeter. ZnO nanotube arrays can be also prepared by coating anodic aluminum oxide (AAO) membranes via atomic layer deposition. However, it yields a relatively low conversion efficiency of 1.6%, primarily due to the modest roughness factor of commercial membranes (Chae et al., 2010).

5.4 Usage of ZnO nanowires as photoanodic material

In 2005, Law et al. first reported the usage of ZnO nanowire arrays in DSSCs by with the intention of replacing the traditional nanoparticle film with a consideration of increasing the electron diffusion length (Law et al., 2007). Nanowires were grown by immersing the seeded substrates in aqueous solutions containing 25 mM zinc nitrate hydrate, 25 mM hexamethylenetetramine, and 5-7 mM polyethylenimine (PEI) at 92 8C for 2.5 h. After this period, the substrates were repeatedly introduced to fresh solution baths in order to obtain continued growth until the desired film thickness was reached. The use of PEI, a cationic polyelectrolyte, is particularly important in this fabrication, as it serves to enhance the anisotropic growth of nanowires. As a result, nanowires synthesized by this method possessed aspect ratios in excess of 125 and densities up to 35 billion wires per square centimeter. The longest arrays reached 20-25 mm with a nanowire diameter that varied from 130 to 200 nm. These arrays featured a surface area up to one-fifth as large as a nanoparticle film.

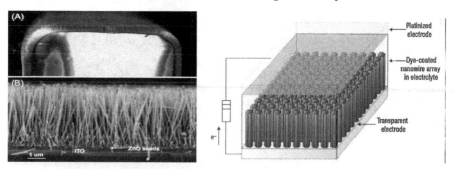

Fig. 6. a) Cross-sectional SEM image of the ZnO-nanowire array and b) Schematic diagram of the ZnO-nanowire dye-sensitized solar cells.

Figure 6a shows a typical SEM cross-section image of an array of ZnO nanowires. It was found that the resistivity values of individual nanowires ranged from 0.3 to 2.0 V cm, with an electron concentration of 1–5 x 10^{18} cm³ and a mobility of 1–5 cm² V¹s¹. Consequently, the electron diffusivity could be calculated as 0.05–0.5 cm² s¹ for a single nanowire. This value is several hundred times larger than the highest reported electron diffusion coefficients for nanoparticle films in a DSSC configuration under operating conditions, that is, 10^{7}–10^{4} cm² s¹ for TiO$_2$ and 10^{5}–10^{3} cm² s¹ for ZnO. A schematic of the construction of DSSC with nanowire array is shown in Figure 6b. Arrays of ZnO nanowires were synthesized in an aqueous solution using a seeded-growth process. This method employed fluorine-doped tin oxide (FTO) substrates that were thoroughly cleaned by acetone/ethanol sonication. A thin film of ZnO quantum dots (dot diameter ~3–4 nm, film thickness ~10–15 nm) was deposited on the substrates via dip coating in a concentrated ethanol solution. For example, at a full sun intensity of 100 x 3mW cm², the highest-surface-area devices with ZnO nanowire arrays were characterized by short-circuit current densities of 5.3–5.85 mA cm², open-circuit voltages of 610–710 mV, fill factors of 0.36–0.38, and overall conversion efficiencies of 1.2–1.5% (Kopidakis et al., 2003).

5.5 Usage of ZnO nanoflowers as photoanodic material

Another interesting morphology is of using ZnO nanoflowers as photoanodic materials for DSSC device fabrication. The shape of nanoflower consists of upstanding stem with irregular branches in all sides of base stem and overall it looks like a flower like morphology. Importance of Nanoflower structure is coverage of ZnO-adsorped dye molecules for effective light harvesting than in in nanorod itself. Because of the fact that nanoflower can be stretch to fill intervals between the nanorods and, therefore, provide both a larger surface area and a direct pathway for electron transport along the channels from the branched "petals" to the nanowire backbone (Fig. 7). ZnO film consisits of nanoflowers can be grown by a hydrothermal method at low temperatures. The typical procedure is as follows: 5 mM zinc chloride aqueous solution with a small amount of ammonia. These as-synthesized nanoflowers, as shown in Figure 7b, have dimensions of about 200 nm in diameter. Then, the ZnO films with "nanoflowers" have been also reported for application in DSSCs. The solar-cell performance of ZnO nanoflower films was characterized by an overall conversion efficiency of 1.9%, a current density of 5.5mA cm², and a fill factor of 0.53. These values are higher than the 1.0%, 4.5 mA cm², and 0.36 for films of nanorod arrays with comparable diameters and array densities that were also fabricated by the hydrothermal method (Jiang et al., 2007).

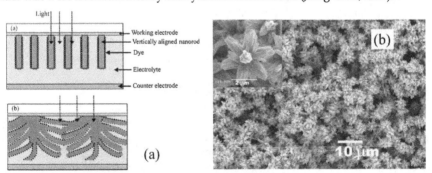

Fig. 7. a) Schematic diagram of the ZnO nanoflower-based dye-sensitized solar cells and b) Top view SEM image of the ZnO-nanoflowers.

5.6 Usage of ZnO nanosheets as photoanodic material

Rehydrothermal growth process of previously hydrothermally grown ZnO nanoparticles can be used to prepare ZnO nanosheets, which are quasi-two-dimensional structures (Suliman et al., 2007, Kakiuchi et al., 2008). Figure 8 shows the SEM images of ZnO nanosheets of low and high magnignified images. ZnO nanosheets are used in a DSSC application, which possess a relatively low conversion efficiency, 1.55%, possibly due to an insufficient internal surface area. It seems that ZnO nanosheetspheres prepared by hydrothermal treatment using oxalic acid as the capping agent may have a significant enhancement in internal surface area, resulting in a conversion efficiency of up to 2.61% (Suliman et al., 2007, Kakiuchi et al., 2008). As for nanosheet-spheres, the performance of the solar cell is also believed to benefit from a high degree of crystallinity and, therefore, low resistance with regards to electron transport.

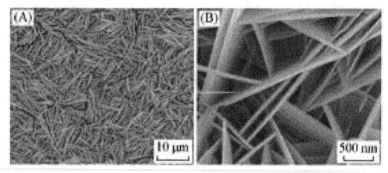

Fig. 8. a) Low and b) High magnified SEM images of the ZnO-nanosheets.

5.7 Usage of ZnO nanobelts as photoanodic material

ZnO nanobelts as photoanodic material can be prepared via an electrodeposition technique. Typically, 1 g of zinc dust mixed with 8 g of NaCl and 4 mL of ethoxylated nonylphenol $[C_9H_{19}C_6H_4(OCH_2CH_2)_nOH]$ and polyethylene glycol $[H(OCH_2CH_2)_nOH]$, and subsequently ground for one hour. The ground paste-like mixture was loaded into an alumina crucible and covered with a platinum sheet leaving an opening for vapor release. The crucible was then loaded into a box furnace and heated at 800°C. Here, ZnO films consists of nanobelt arrays as shown in Figure 9a and it also proposed to use for DSSC applications. In fabricating these nanobelts, polyoxyethylene cetylether was added in the electrolyte as a surfactant. The ZnO nanobelt array obtained shows a highly porous stripe structure with a nanobelt thickness of 5 nm, a typical surface area of 70 $m^2 g^1$, and a photovoltaic efficiency as high as 2.6%.

5.8 Usage of ZnO nanotetrapods as photoanodic material

A three-dimensional structure of ZnO tetrapod that consisting of four arms extending from a common core, as showin in Figure 9b (Jiang et al., 2007 & Chen et al., 2009). The length of the arms can be adjusted within the range of 1–20 mm, while the diameter can be tuned from 100 nm to 2 mm by changing the substrate temperature and oxygen partial pressure during vapor deposition. Multiple-layer deposition can result in tetrapods connected to each other so as to form a porous network with a large specific surface area. The films with ZnO tetrapods used in DSSCs have achieved overall conversion efficiencies of 1.20– 3.27%. It was

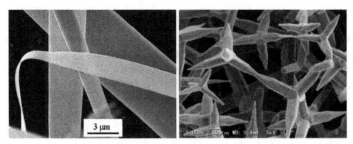

Fig. 9. SEM images of a) ZnO-nanobelt and b) ZnO nanotetrapods.

reported that the internal surface area of tetrapod films could be further increased by incorporating ZnO nanoparticles with these films, leading to significant improvement in the solar-cell performance. Another type of nanomaterials such as nanoporous film also leads to have the maximum coversion efficiency of 4.1% with N719 dye (Hosono et al., 2005).

5.9 Usage of ZnO aggregates as photoanodic material

So far, the maximum overall energy conversion efficiency was reported up to 5.4% from the ZnO film consisits of polydisperse ZnO aggregates, when compared to other nanostructures conversion efficiency of 1.5–2.4% for ZnO nanocrystalline films, 0.5–1.5% for ZnO Nanowire films, and 2.7–3.5% for uniform ZnO aggregate films (Desilvestro et al., 1985, Chou et al., 2007 & Zhang et al., 2008). The overall conversion efficiency of 5.4% with a maximum short-circuit current density of 19mA cm^2 are observed. In other words, the aggregation of ZnO nanocrystallites is favorable for achieving a DSSC with high performance, as shown in Figure 10. This result definitely shock us, since, many gourps were seriously working in synthesizing nanostructured material for DSSC. Here, though the ZnO aggregates are falls in submicron range, individual ZnO nanoparticles are in less than 20 nm. In Figure 10, the film is well packed by ZnO aggregates with a highly disordered stacking, while the spherical aggregates are formed by numerous interconnected nanocrystallites that have sizes ranging from several tens to several hundreds of nanometers. The preparation of these ZnO aggregates can be achieved by hydrolysis of zinc salt in a polyol medium at 160 °C (Chou et al., 2007). By adjusting the heating rate during synthesis and using a stock solution containing ZnO nanoparticles of 5 nm in diameter, ZnO aggregates with either a monodisperse or polydisperse size distribution can be prepared (Zhang et al., 2008).

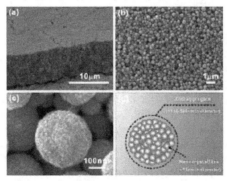

Fig. 10. SEM images of ZnO film with aggregates synthesized at 160 °C and a schematic showing the structure of individual aggregates.

6. Limitation on ZnO-based DSSCs

Although ZnO possesses high electron mobility, low combination rate, good crystallization into an abundance of nanostructures and almost an equal band gap and band position as TiO_2, the photoconversion efficiency of ZnO based DSSC still limited. The major reason for the lower performance in ZnO-based DSSCs may be explained by the a) formation of $Zn^2þ$/dye complex in acidic dye and b) the slow electron-injection flow from dye to ZnO. $Zn^2þ$/dye complex formation mainly occurs while ZnO is dipped inside the acidic dye solution for the dye adsorption for a long time. Ru based dye molecules consisits of carboxylic functional group for coordination, dye solution mostly existing in acidic medium. Therefore, the $Zn^2þ$/dye complex is inevitable. The formation of $Zn^2þ$/dye complex has been attributed to the dissolution of surface Zn atoms by the protons released from the dye molecules in an ethanolic solution. For lower electron-injection efficiency is reported of using ZnO material with Ru-based dyes when compared to TiO_2. In ZnO, the electron injection is dominated by slow components, whereas for TiO_2 it is dominated by fast components, leading to a difference of more than 100 times in the injection rate constant. For example, either ZnO or TiO_2, the injection of electrons from Ru-based dyes to a semiconductor shows similar kinetics that include a fast component of less than 100 fs and slower components on a picosecond time scale (Anderson et al., 2003). That is, the ZnO conduction bands are largely derived from the empty s and p orbitals of $Zn^2þ$, while the TiO2 conduction band is comprised primarily of empty 3d orbitals from $Ti^4þ$ (Anderson et al., 2004). The difference in band structure results in a different density of states and, possibly, different electronic coupling strengths with the adsorbate.

7. Alternative dyes for ZnO

According to the limitations of ZnO based DSSC, the lower electron injection and the instability of ZnO in acidic dyes, the alternative type dyes will provide a new pathway for useage of ZnO nanomaterials as photoanodic materials for effective solar power conversion. The list of other alternative dyes were compiled and given in Table 4. The new types of dyes should overcome above mentioned two different limitations and it should be chemically bonded to the ZnO semiconductor for effective for light absorption in a broad wavelength range. Already few research groups were already developed with the aim of fulfilling these criteria. The various new types of dyes include heptamethine-cyanine dyes adsorbed on ZnO for absorption in the red/near-infrared (IR) region (Matsui et al., 2005 & Otsuka et al., 2006 & 2008), and unsymmetrical squaraine dyes with deoxycholic acid, which increases photovoltage and photocurrent by suppressing electron back transport (Hara et al., 2008). Mercurochrome ($C_{20}H_8Br_2HgNa_2O$) is one of the newly developed photosensitizers that, to date, is most suitable for ZnO, offering an IPCE as high as 69% at 510 nm and an overall conversion efficiency of 2.5% (Hara et al., 2008 & Hosono et al., 2004). It was also reported that mercurochrome photosensitizer could provide ZnO DSSCs with a fill factor significantly larger than that obtained with N3 dye, where the latter device was believed to possess a higher degree of interfacial electron recombination due to the higher surface-trap density in the N3-dye-adsorbed ZnO. Eosin Y is also a very efficient dye for ZnO-based DSSCs, with 1.11% conversion efficiency for nanocrystalline films (Rani et al., 2008). When eosin Y is combined with a nanoporous film, overall conversion efficiencies of 2.0–2.4% have been obtained (Hosono et al., 2004 & Lee et al., 2004). Recently, Senevirathne et al. reported that the use of acriflavine (1,6diamino-10-methylacridinium chloride) as a photosensitizer for ZnO could generate photocurrents that are an order of magnitude higher than in the case of TiO_2 (Senevirathne et al., 2008). Three triphenylamine dyes based on low-cost methylthiophene as the p-conjugated spacer were designed and synthesized as the dye sensitizers for DSSCs applications. The high photovoltaic performances of the DSSCs based on these as-synthesized dyes were obtained. Though the introduction of vinyl unit in the p-conjugated spacer can obtainred-shifted absorption spectra, it does not give a positive effect on the photovoltaic performance of the DSSCs due to unfavorable back-electron transfer and decrease of the open-circuit voltage. On the basis of optimized conditions, the DSSCs based on these three as-synthesized dyes exhibited the efficiencies ranging from 7.83% to 8.27%, which reached 80 to 85% with respect to that of an N719-based device. The high conversion efficiency and easy availability of rawmaterials reveal thatthese metal-free organic dyes are promising in the development of DSSCs (Tian et al., 2010).

Structure	Photosensitizer	Efficiency
Nanoparticles	heptamethine cyanine	0.16%, 0.67
	unsymmetrical squaraine	1.5%
	eosin-Y	1.11%
	acriflavine	0.588%
	mercurochrome	2.5%
Nanoporous films	D149	4.27%
	eosin-Y	2.0%, 2.4%
	eosin-Y	3.31%(0.1 sun)
Nanowires	QDs (CdSe)	0.4%

Table 4. The list of other alternative dyes for ZnO based DSSC.

8. Conclusions

ZnO is believed to be a superior alternative material to replace the existing TiO_2 photoanodic materials used in DSSC and has been intensively explored in the past decade due to its wide band gap and similar energy levels to TiO_2. More important, its much higher carrier mobility is favorable for the collection of photoinduced electrons and thus reduces the recombination of electrons with tri-iodide. Although the formation of $Zn^2\flat/dye$ complex is inevitable due to the dissolution of surface Zn atoms by the protons released from the dye molecules in an ethanolic solution, selection of other alternative dye molecules will definitely helps to boost the conversion efficiency to much higher level. Therefore, the recent development on the synthesis of metal-free dye molecules will lead the DSSC device fabrication to the new height as for the cost effectiveness and simple technique are concern.

9. References

Liu, J. (2008). Oriented Nanostructures for Energy Conversion and Storage. Chem. *Sus. Chem.*, 1, pp. 676.

Bagnall, D. M. (2008). *Photovoltaic technologies*. Energ. Pol., 36, pp. 4390.

Chapin D. M. (1954). A new silicon p-n junction photocell for converting solar radiation into electrical power. *J. Appl. Phys.* 25, pp. 676.

Afzaal, M. (2006). Recent developments in II–VI and III–VI semiconductors and their applications in solar cells. *J. Mater. Chem.* 16, pp. 1597.

Birkmire, R. W. (2001). Compound polycrystalline solar cells: Recent progress and Y2 K perspective. *Sol. Energ. Mat. Sol.* 65, pp. 17.

Gratzel, M. (2001). Review article Photoelectrochemical cells. *Nature,* 414, pp. 338.

Gratzel, M. (2000). Perspectives for dye-sensitized nanocrystalline solar cells. *Prog. Photovoltaics,* 8, pp. 171.

Gratzel, M. (2003). Dye-sensitized solar cells. *J. Photochem. Photobiol. C.* 4, pp. 145.

Gratzel, M. (2007). Photovoltaic and photoelectrochemical conversion of solar energy. *Phil. Trans. Math. Phys. Eng. Sci.* 365, pp. 993.

Nelson, J. (2004). Random walk models of charge transfer and transport in dye sensitized systems. *Coord. Chem. Rev.* 248, pp. 1181.

Fujishima, A. (1972). Electrochemical photolysis of water at a semiconductor electrode. *Nature,* 38, pp. 5358.

Desilvestro, J.; Gratzel, M.; Kavan, L.; Moser, J. & Augustynski, J. (1985). Highly Efficient Sensitization of Titanium Dioxide. *Journal of the American Chemical Society,* 107, 10, pp. 2988-2990.

Nazeerudin, M. K. (2003). Conversion of light to electricity by cis-X2bis(2,2'-bipyridyl-4,4'-dicarboxylate)ruthenium(II) charge-transfer sensitizers (X = Cl-, Br-, I-, CN-, and SCN-) on nanocrystalline titanium dioxide electrodes *Journal.of the American Chemical Society* 115, 14, pp. 6382.

Chiba, Y. (2006). Conversion efficiency of 10.8% by a dye-sensitized solar cell using a TiO2 electrode with high haze. *Applied Physics Letters,* 88, 22 pp. 223505.

Tornow, J. Transient Electrical Response of Dye-Sensitized ZnO Nanorod Solar Cells. *J. Phys. Chem. C.* 111, pp. 8692.

Djurisic, A.B. (2006). Optical Properties of ZnO Nanostructures. *Small.* 2, pp. 944.

Tornow, J. (2008). Voltage bias dependency of the space charge capacitance of wet chemically grown ZnO nanorods employed in a dye sensitized photovoltaic cell. *Thin Solid Films.* 516, pp. 7139.

Bittkau, K. (2007). Near-field study of optical modes in randomly textured ZnO thin films. *Superlatt. Microstuct.* 42, pp. 47.

Xi, Y.Y. (2008). Electrochemical Synthesis of ZnO Nanoporous Films at Low Temperature and Their Application in Dye-Sensitized Solar Cells . *J. Electrochem. Soc.* 155, pp. D595.

Jagadish, C. (2006). Zinc oxide bulk, thin films and nanostructures: processing, properties and applications, Publisher: Elsevier Science, 1 edition, pp 3.

Dulub, O. (2002). STM study of the geometric and electronic structure of ZnO(0001)-Zn, (0001)-O, (1010), and (1120) surfaces, *Surf. Sci.* 519, pp. 201-217.

Wander, A. (2001). Stability of polar oxide surfaces, *Phys.Rev. Lett.* 86, pp. 3811-3814.

Staemmler, V. (2003). Stabilization of polar ZnO surfaces: validating microscopic models by using CO as a probe molecule, *Phys. Rev. Lett.* 90, pp. 106102-1-4.

Singh, S. (2007). Structure, microstructure and physical properties of ZnO based materials in various forms: bulk, thin film and nano, *J. Phys. D: Appl. Phys.* 40, pp. 6312-6327.

Pearton, S.J. (2005). Recent progress in processing and properties of ZnO, *Progress in Materials Science*. 50, pp. 293- 340.

Florescu, D. (2002). High spatial resolution thermal conductivity of bulk ZnO (0001), *J Appl Phys*. 91, pp. 890-892.

Hosokawa, M. (2007). Nanoparticle Technology Handbook Elsevier, Amsterdam.

Lee, J. S. (2003). ZnO nanomaterials synthesized from thermal evaporation of ballmilled ZnO powders, *J. Cryst. Growth*. 254, pp. 423-431.

Zhao, Q. X. (2007). Growth of ZnO nanostructures by vapor–liquid–solid method, *Appl. Phys. A*, 88, pp. 27-30.

Huang, M. H. (2001). Catalytic growth of zinc oxide nanowires by vapor transport, *Adv. Mater*. 13, pp. 113- 116.

Sun, Y. (2004). Growth of aligned ZnO nanorod arrays by catalyst-free pulsed laser deposition methods, *Chem. Phys. Lett*. 39, pp. 621-26.

Wu, J . (2002). Low-temperature growth of well-aligned ZnO nanorods by chemical vapor deposition, *Adv .Mater*. 14, pp. 215-218.

Park, W. I. (2002). Metalorganic vapor-phase epitaxial growth of vertically well-aligned ZnO nanorods, *Appl. Phys. Lett*. 80, pp. 4232-4234.

Yu, H. D. (2005). A general lowtemperature route for large-scale fabrication of highly oriented ZnO nanorod/nanotube arrays. *J. Am. Chem. Soc*. 127, pp. 2378-2379.

Zeng, L.Y. (2006). Dye-Sensitized Solar Cells Based on ZnO Films. *Plasma Sci. Tech*. 2006, 8, pp. 172.

Keis, K. (2000). Studies of the Adsorption Process of Ru Complexes in Nanoporous ZnO Electrodes. *Langmuir*. 16, pp. 4688.

Suliman, A.E.(2007). Preparation of ZnO nanoparticles and nanosheets and their application to dye-sensitized solar cells Sol. *Energ. Mat. Sol. Cells*. 91, pp. 1658.

Gonzalez-Valls, I. (2010). Dye sensitized solar cells based on vertically-aligned ZnO nanorods: effect of UV light on power conversion efficiency and lifetime. *Energy Environ. Sci.*, 3, pp. 789-795.

Lai, M. H. (2010). ZnO-Nanorod Dye-Sensitized Solar Cells: New Structure without a Transparent Conducting Oxide Layer. *International Journal of Photoenergy*, pp. Article ID 497095, 5 pages

Hsu, Y.F. (2008). ZnO nanorods for solar cells: Hydrothermal growth versus vapor deposition, *Appl. Phys. Lett*. 92, pp. 133507.

Chen, H.H. (2008). Dye-sensitized solar cells using ZnO nanotips and Ga-doped ZnO films. *Semicond. Sci. Technol*. 23, pp. 045004.

Martinson, A.B.F. (2007). ZnO Nanotube Based Dye-Sensitized Solar Cells. *Nano Lett*. 7, pp. 2183.

Lin, C. F. (2008). Electrodeposition preparation of ZnO nanobelt array films and application to dye-sensitized solar cells. J. *Alloys Compd*. 462, pp. 175.

Kakiuchi, K. (2008). Fabrication of ZnO films consisting of densely accumulated mesoporous nanosheets and their dye-sensitized solar cell performance. *Thin Solid Films*. 516, pp. 2026.

Chen, W. (2009). A new photoanode architecture of dye sensitized solar cell based on ZnO nanotetrapods with no need for calcination. *Electrochem. Commun.* 11, pp. 1057.

Jiang, C. Y. (2007). Improved dye-sensitized solar cells with a ZnO-nanoflower photoanode. *Appl. Phy. Lett.* 90, pp. 263501.

Chen, Z.G. (2006). Electrodeposited nanoporous ZnO films exhibiting enhanced performance in dye-sensitized solar cells. *Electrochim. Acta.* 51, pp. 5870.

Hosono, E. (2005). The Fabrication of an Upright-Standing Zinc Oxide Nanosheet for Use in Dye-Sensitized Solar Cells. *Adv. Mater.* 17, pp. 2091.

Kakiuchi, K. (2006). Enhanced photoelectrochemical performance of ZnO electrodes sensitized with N-719. *J. Photochem. Photobiol. A* 179, pp. 81.

Guo, M. (2005). Hydrothermal growth of perpendicularly oriented ZnO nanorod array film and its photoelectrochemical properties. *Appl. Surf. Sci.* 249, pp. 71.

Guo, M. (2005). The effect of hydrothermal growth temperature on preparation and photoelectrochemical performance of ZnO nanorod array films. *J. Solid State Chem.* 178, pp. 3210.

Rao, A. R. (2008). Achievement of 4.7% conversion efficiency in ZnO dye-sensitized solar cells fabricated by spray deposition using hydrothermally synthesized nanoparticles. *Nanotechnology.* 19, pp. 445712.

Wu, J. J. (2007). Effect of dye adsorption on the electron transport properties in ZnO-nanowire dye-senstized solar cells. *Appl. Phy. Lett.* 2007, 90, pp. 213109.

Law, M. (2005). Nanowire dye-sensitized solar cells. *Nat. Mater.* 4, pp. 455.

Zhang, R. (2008). High-Density Vertically Aligned ZnO Rods with a Multistage Terrace Structure and Their Improved Solar Cell Efficiency. *Cryst. Growth Des.* 8, pp. 381.

Chou, T.P. (2007). Hierarchically Structured ZnO Film for Dye-Sensitized Solar Cells with Enhanced Energy Conversion Efficienc. *Adv. Mater.* 19, pp. 2588.

Zhang, Q.F. (2008). Polydisperse aggregates of ZnO nanocrystallites: a method for energy-conversion-efficiency enhancement in dye-sensitized solar cells, *Adv. Funct. Mater.* 18, pp. 1654.

Chou, T.P. (2007). Effects of Dye Loading Conditions on the Energy Conversion Efficiency of ZnO and TiO$_2$ Dye-Sensitized Solar Cells. *Phys. Chem. C.* 111, pp.18804.

Uthirakumar, p. (2006). ZnO nanoballs synthesized from a single molecular precursor via non-hydrolytic solution route without assistance of base, surfactant, and template etc . *Phys. Lett. A,* 359, pp. 223.

Uthirakumar, p. (2007). Nanocrystalline ZnO particles: Low-temperature solution approach from a single molecular precursor. *J. Cryst. Growth.* 304, pp. 150.

Uthirakumar, p. (2009). Zinc Oxide Nanostructures Derived from a Simple Solution Method for Solar Cells and LEDs. *Chemical Engineering Journal,* 155, pp. 910-915.

Uthirakumar, p. (2009). Effect of Annealing Temperature and pH on the Morphology and Optical Properties of Highly Dispersible ZnO nanoparticles. *Material Characterization,* 60(11), pp. 1305-1310.

Eom, S. H. (2008). Preparation and characterization of nano-scale ZnO as a buffer layer for inkjet printing of silver cathode in polymer solar cells. *Sol. Energy Mater. Sol. Cells*, 92, pp. 564.

Hosono, E. (2004). Growth of layered basic zinc acetate in methanolic solutions and its pyrolytic transformation into porous zinc oxide films. *J. Colloid Interface Sci.* 272, 391.

Hosono, E. (2008). Metal-free organic dye sensitized solar cell based on perpendicular zinc oxide nanosheet thick films with high conversion efficiency. *Dalton Trans.* pp. 5439.

Charoensirithavorn, P. (2006). Dye-sensitized Solar Cell Based on ZnO Nanorod Arrays. *The 2nd Joint International Conference on Sustainable Energy and Environment.* 21-23 November 2006, Bangkok, Thailand.

Han, J. (2010). ZnO nanotube-based dye-sensitized solar cell and its application in self-powered devices, *Nanotechnology* 21, pp. 405203.

Martinson, A.B.F. (2007). ZnO Nanotube Based Dye- ensitized Solar Cells. *Nano letters.* 7, pp. 2183.

Chae, K. W. (2010). Low-temperature solution growth of ZnO nanotube arrays, *Beilstein J. Nanotechnol.* 1,pp. 128–134.

Kopidakis, N. (2003). Transport-Limited Recombination of Photocarriers in Dye-Sensitized Nanocrystalline TiO$_2$ Solar Cells. *J. Phys. Chem. B.* 107, pp. 11307.

Zhang, Q. (2010). Synthesis of ZnO Aggregates and Their Application in Dye-sensitized Solar Cells. *Material Matters. 5.2*, pp. 32.

Anderson, N.A. (2003). X. Ai,T.Q. Lian, Electron Injection Dynamics from Ru Polypyridyl Complexes to ZnO Nanocrystalline Thin Films. *J. Phys. Chem. B* 107, pp. 14414.

Anderson, N.A. (2004). Ultrafast electron injection from metal polypyridyl complexes to metal-oxide nanocrystalline thin films. *Coord. Chem. Rev.* 248, pp. 1231.

Matsui, M. (2005). Application of near-infrared absorbing heptamethine cyanine dyes as sensitizers for zinc oxide solar cell . *Synth. Met.* 148, pp. 147.

Otsuka, A. (2008). Simple Oligothiophene-Based Dyes for Dye-Sensitized Solar Cells (DSSCs): Anchoring Group Effects on Molecular Properties and Solar Cell Performance. *Chem. Lett.* 37, pp. 176.

Otsuka, A. (2006). Dye Sensitization of ZnO by Unsymmetrical Squaraine Dyes Suppressing Aggregation. *Chem. Lett.* 35, pp. 666.

Hara, K. (2003). Molecular Design of Coumarin Dyes for Efficient Dye-Sensitized Solar Cells. *Phys. Chem. B*, 107 (2), pp 597–606.

Hara, K. (2000). Highly efficient photon-to-electron conversion with mercurochrome-sensitized nanoporous oxide semiconductor solar cells. *Sol. Energ. Mat. Sol. Cells.* 64, pp. 115.

Hosono, (2004). Synthesis, structure and photoelectrochemical performance of micro/nano-textured ZnO/eosin Y electrodes. *Electrochim. Acta.* 49 (14), pp. 2287.

Rani, S. (2008). Synthesis of nanocrystalline ZnO powder via sol–gel route for dye-sensitized solar cells. *Sol. Energ. Mat. Sol. Cells.* 92, pp. 1639.

Lee, W. J. (2004). Fabrication and Characterization of Eosin-Y-Sensitized ZnO Solar Cell, *Jpn. J. Appl. Phys.* 43, pp. 152.

Senevirathne, M.K.I. (2008). Sensitization of TiO_2 and ZnO nanocrystalline films with acriflavine. *Photochem. Photobiol. A.* 195, pp. 364.

Tian, Z. (2010). Low- cost dyes based on methylthiophene for high-performance dye-sensitized solarcells. *Dyes and Pigments* 87, pp. 181-187.

10

Dye Sensitized Solar Cells as an Alternative Approach to the Conventional Photovoltaic Technology Based on Silicon - Recent Developments in the Field and Large Scale Applications

Elias Stathatos
Technological-Educational Institute of Patras,
Electrical Engineering Department, Patras,
Greece

1. Introduction

Utilization of renewable energies is of major importance because of the increase in fossil energy costs in combination with carbon dioxide reduction preventing global warming. The importance of the solar energy can be considered as the sustainable energy which may successfully satisfy a part of the energy demand of future generations. The 3×10^{24} joule/year energy supply from sun to the earth is ten thousand times more than the global need. It means that the use of 10% efficiency photovoltaic cells could cover the present needs in electricity covering only the 0.1% of earth's surface (Wu, et al. 2008). Handling this opportunity of solar energy utilization is a big bet for the future. Besides the development of new clean techniques to the electrical power generation is urgently important in order to protect global environment and assure economic growth of sustainable resources. Taking into account the present status in photovoltaic technology, some improvements have to be made which are summarized in three basic fields: (a) costs, (b) in their applicability and (c) sustainability. Although the cost per peak watt of crystalline silicon solar cells has significantly dropped, it is still expensive compared to the conventional grid electricity resources. Silicon wafers made of pure semiconducting material to avoid limitations in energy conversion, are still expensive. For this reason developments on potentially cheaper solar cells based on thin-film technology have been made. According to this technology, thin films made of purely inorganic materials such as amorphous silicon, cadmium telluride, and copper indium diselenide successfully prepared on glass substrates.

Almost two decades ago, dye sensitized solar cells (DSSCs) were proposed as low cost alternatives to the conventional amorphous silicon solar cells, owing to the simplicity of their fabrication procedures, practically under ambient conditions with mild chemical processes. DSSCs are placed in the category of third generation photovoltaics where new trends in the photovoltaic technology are applied. In the 1st generation PV cells, the electric interface is made between doped n-type and p-type bulk silicon. 1st generation PV cells provide the highest so far conversion efficiency. The 2nd generation PV cells are based on thin film technology. These cells utilize less material and they thus drop the production cost, however, they are less efficient than the bulk cells. Both 1st and 2nd generation cells are based on opaque materials and necessitate front-face illumination and moving supports to follow sun's position. Thus they may be either set up in PV parks or on building roofs. 3rd generation solar cells, are based on nanostructured (mesoscopic) materials and they are made of purely organic or a mixture of organic and inorganic components, thus allowing for a vast and inexhaustible choice of materials. Because of their mesoscopic character, it is possible to make transparent cells, which can be used as photovoltaic windows. Photovoltaic windows can be functioned by front-face light incidence but also by diffuse light and even by back face light incidence. Also because of their mesoscopic nature, 3rd generation solar cells are easy to make at ambient conditions, not necessitating severe measures of purity, thus dropping production cost. Among the different possibilities of 3rd generation solar cells, DSSC have the most promising prospect. The overall efficiency of ~12% (in laboratory and small size cells) placed DSSCs as potential inexpensive alternatives to solid state devices. Since the pioneer work of M. Grätzel and co-workers an intense interest to the development of such kind of solar cells has been recorded because of their low cost, simple preparation procedures and benign methods of construction compared with conventional methods applied in first and second generation photovoltaic technology (O'Regan & Grätzel, 1991). Although the solar to electrical energy conversion efficiencies recorded for DSSCs are lower than those measured for silicon based solar cells, a high potential for improvement in their efficiency, stability and commercialization has been announced till nowadays (Grätzel, 2006; Goldstein et al., 2010; Hinsch et al., 2009).

2. Principles of operation and cell structure

The working principle of a DSSC substantially differs from that of a conventional solar cell based on silicon. In silicon solar cell a *p-n* junction by joining semiconductors of different

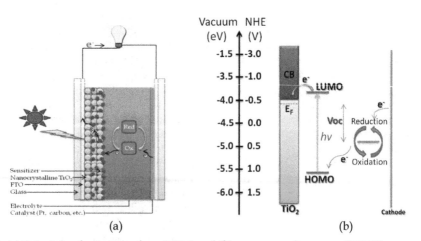

Fig. 1. (a) Principle of operation for a DSSC and (b) an energy diagram of DSSC's operation.

charge carriers' concentration in a very close contact is necessary. In this case the processes of light absorption and charge transport are caused in the same material. In DSSCs, these fundamental processes are occurred in different materials which avoid the premature recombination of electrons and holes. As these processes do not happen at the same material ultrapure materials are not required for a high performance DSSC. DSSCs are composed of four major components: a nanostructured n-type semiconductor, typically TiO_2, a dye-sensitizer to absorb visible light, an electrolyte, which creates the interface with the semiconductor and a counter electrode carrying an electrocatalyst, which facilitates transfer of electrons to the electrolyte. Figure 1a illustrates the basic principle of cell operation while Figure 1b the energy diagram of basic components of the DSSC.

Charge separation is occurred by the different electrochemical potentials between different species such as negative electrode (TiO_2/sensitizer) and electrolyte. Any electrostatic potential like in the case of silicon based solar cells is then ignored when a minimum concentration of 0.4M of mobile ions exist in the electrolyte (Grätzel & Durrant, 2008). The semiconductor must provide large active interface both for the attachment of the sensitizer and the contact with the electrolyte. Therefore, the semiconductor can be only conceived in nanostructured form. The sensitizer must have a large extinction coefficient and its energy states must match with those of the semiconductor so as to allow extensive light absorption and efficient excited–electron injection into the conduction band of the semiconductor. The electrolyte must have appropriate electrochemical potential so as to combine with the semiconductor and to efficiently provide charge mobility in a cyclic manner. The dye is regenerated by electrons donated from the electrolyte. The iodide is then regenerated by the reduction of triiodide at the positive electrode, and the circuit is completed by the electron migration through the external circuit. Finally, the counter electrode must efficiently catalyze the transfer of electrons from the external circuit to the liquid phase, i.e. the electrolyte. The open circuit voltage of the cell generated under illumination is attributed to the difference between the Fermi level of the nanostructured semiconductor and the electrochemical potential of the electrolyte. The photoelectrochemical processes occur in a DSSC can be expressed in equations 1-6 (Wu et al., 2008).

$$TiO_2|S + hv \rightarrow TiO_2|S^* \quad \text{(dye excitation)} \tag{1}$$

$$TiO_2|S^* \rightarrow TiO_2|S^+ + e^-_{(CB)} \quad \text{(electron injection in ps scale)} \tag{2}$$

$$TiO_2|S^* + 3I^- \rightarrow TiO_2|S + I_3^- \quad \text{(dye regeneration in μs scale)} \tag{3}$$

$$I_3^- + 2e^-_{(Pt)} \rightarrow 3I^- \quad \text{(reduction)} \tag{4}$$

While the dark reactions which may also happen are:

$$I_3^- + 2e^-_{(CB)} \rightarrow 3I^- \quad \text{(recombination to electrolyte from ms to s scale)} \tag{5}$$

$$TiO_2|S^+ + e^-_{(CB)} \rightarrow TiO_2|S \quad \text{(recombination from μs to ms scale)} \tag{6}$$

From equations described above it is obvious that several issues have to be simultaneously satisfied in order to achieve an efficient solar cell based on nanostructured dye sensitized semiconductors. As a first issue we may refer that the dye has to be rapidly reduced to its ground state after it is oxidized while the electrons are injected into the conduction band of the TiO_2 otherwise the solar cell performance will be low. This means that the chemical potential of the iodide/triiodide redox electrolyte should be positioned in more negative values than the oxidised form of the dye. Furthermore the nanocrystalline TiO_2 film must be able to permit fast diffusion of charge carriers to the conductive substrate and then to external circuit avoiding recombination losses, while good interfacial contact between electrolyte and semiconductor has to be ensured (Bisquert et al., 2004). Electrolyte long term stability (chemical, thermal, optical) which will guarantee solar cell high performance is under continuous consideration as in common DSSC structures the electrolyte is in the form of a volatile liquid bringing out the obvious problem of sealing (Zhang et al., 2011). Finally, the optimized concentration of redox couple for the cell efficiency has to satisfy one more parameter of the optical transparency in the visible region otherwise the absorbed light from the dye will be minimized and also triiodide can react with injected electrons increasing the dark current of the cell.

Although, the charge transport rate in DSSCs is relatively slow compared with conventional photovoltaics and the interface where the charge carrier could recombine is wide. Because of the mesoporous structure the charge collection quantum efficiency is surprisingly close to unity (Grätzel & Durrant, 2008). This is caused because of the slow rate constant for the interfacial charge recombination of injected electrons with the oxidised redox couple. The presence of a suitable catalyst (e.g. Pt) raises an activation barrier in one of the intermediate steps of redox reactions resulting in a slow overall rate constant for this reaction. This low rate constant for this recombination reaction on TiO_2, affect to an increased efficiency for DSSCs.

The kinetic competition between charge transport and recombination in DSSCs can be analysed in terms of an effective carrier diffusion length L_n, given by $L_n = [D_{eff} \tau]^{1/2}$ where D_{eff} is the effective electron diffusion length, and τ the electron lifetime due to the charge-recombination reaction given by eq. 5 (Peter & Wijayantha, 2000). D_{eff} strongly depends on the position of the quasi Fermi level in the semiconductor and therefore on the light intensity. Typical values at 1 sun are $1.5 \ 10^{-5} \ cm^2s^{-1}$. Since diffusion is the only driving force for electron transport, the diffusion length DL must be at least as long a the thickness of the TiO_2 electrode. D_{eff} generally increases with light intensity while τ proportionally decreases. As a consequence the diffusion length is independent of the light intensity. Typical values for diffusion length are 5–20 µm. These limitations set the rules according to which the researchers are challenged to make a choice of materials that will lead to efficient cell functioning.

3. DSSCs' basic components

The basic structure of a DSSC, as it is referred in previous section, is consisted of two glass electrodes in a sandwich configuration. For the first electrode (negative) a nanocrystalline n-type semiconductor, typically titanium dioxide film is deposited on a transparent conductive glass (TCO) (Fig.2) and then a dye-sensitizer is adsorbed and chemically anchored in order to sensitize the semiconductor in the visible. For this purpose, the dye sensitizer bears carboxylate or phosphonate groups, which interact with surface –OH groups on the titanium dioxide. Several efforts have been made to apply dyes of various structures; however, Ru-bipyridine complexes have established themselves as choice sensitizers (Xia & Yanagida, 2009). This is the negative electrode of the solar cell. A similar transparent conductive glass (positive electrode) covered with a thin layer of platinum is faced to the previous electrode. The space between the two electrodes is filled with an electrolyte. The most efficient electrolytes applied with DSSCs are liquid electrolytes with dissolved I^-/I_3^- redox couple, which obtained by co-dissolving an iodide salt with iodine (Hagfeldt & Grätzel, 2000). Since some crystallization problems have been encountered with simple salts, like LiI or KI, recent research is concentrated on the employment of ionic liquids, principally, alkylimidazolium iodides (Papageorgiou et al., 1996).

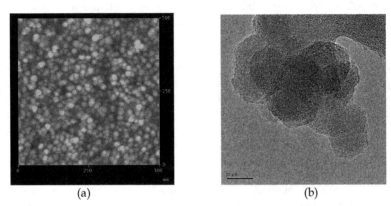

(a)   (b)

Fig. 2. AFM (a) and HR-TEM (b) images of a nanocrystalline TiO_2 film.

3.1 Nanocrystalline semiconductor

In DSSC technology a variety of nanocrystalline mesoporous metal oxides have been used such as TiO_2, ZnO, SnO_2 and Nb_2O_5 (Sayama, et al., 1998, Jose, et al., 2009). Despite the fact that some of them exhibited promising results in cells' performance only titanium dioxide has extensively used because of some advantages which are only present in this oxide. TiO_2 performs excellent thermal stability; it is impervious to chemicals and non-toxic and finally a cheap material. The common crystalline form in application to solar cells is the anatase although a mixture of anatase/rutile form is often used mainly by the formation of very active commercial Degussa-P25 powder. Rutile has proved to be less active as it is less chemical stable than anatase form. Combinations of metal oxides as negative electrodes have also been examined such as WO_3/TiO_2, TiO_2/ZrO_2 and SnO_2/ZnO or SnO_2/TiO_2 with moderate results (Tennakone, et al., 1999). In the case of mixed oxides, the core-shell nanostructure formation is mentioned as a new class of combinational system which is typically comprised of a core made of nanomaterials and a shell of coating layer covering on the surface of core nanomaterials (Zhang & Cao, 2011). The use of core-shell nanostructures is usually refereed to lower the charge recombination in the TiO_2 nanoparticles and it is based on the hypothesis that a coating layer may build up an energy barrier at the semiconductor/electrolyte interface retarding the reaction between the photogenerated electrons and the redox species in electrolyte. Different systems that consisted of mesoporous TiO_2 films coated with oxides such as Nb_2O_5, ZnO, $SrTiO_3$, ZrO_2, Al_2O_3 and SnO_2 are also referred. The results revealed that, compared to photoelectrode made of bare TiO_2 nanoparticles, the use of e.g. Nb_2O_5 shell might increase both the open circuit voltage and the short circuit current of the cells.

The basic goal in films preparation is the high surface area of the inorganic semiconductor particles in order to achieve high amounts of dye adsorbed on it. Therefore, a much interest has been drawn to the preparation of highly crystalline mesoporous materials in the form of homogeneous films with an average thickness of 6-12 μm. Usually TiO_2 nanoparticles are fabricated by the aqueous hydrolysis of a titanium alkoxide precursor. It is then followed by autoclaving at temperatures up to 240°C to achieve the desired nanoparticle size and crystallinity (anatase) (Barbe et al., 1997). The nanoparticles are deposited as a colloidal suspension by screen printing or by spreading with a doctor blade technique, followed by sintering at ~450°C to achieve good interparticle connections. The film porosity is maintained by the addition of surfactants or organic fillers; the organic content is removed after sintering of the films in order to obtain pure titanium dioxide (Stathatos et al., 2004). Figure 3 shows a SEM cross sectional image of a mesoporous TiO_2 film prepared by titanium dioxide powder formed with screen printing method. The average pore size is 15 nm and particle diameter 20-25 nm. Film morphology is a crucial parameter in DSSCs' performance mainly to the influence in electron recombination rate. As referred in literature this phenomenon usually happens in the contact between TiO_2 film and conductive substrate (Zhu, et al., 2002). Therefore, a condensed non-porous thin film of TiO_2 is formed between nanocrystalline thick film and TCO substrate and referred as "blocking layer". The thickness of the compact film is around a few hundreds of nanometres. An alternative method to prepare highly porous nanocrystalline TiO_2 with even more smaller particles is the sol-gel. The sol-gel method for the synthesis of inorganic or nanocomposite organic/inorganic gels has become one of the most popular chemical procedures (Stathatos et al., 1997). This popularity stems from the fact that sol-gel synthesis is easy and it is carried out at ambient or slightly elevated temperatures so that it allows non-destructive organic doping (Brinker & Scherer, 1990).

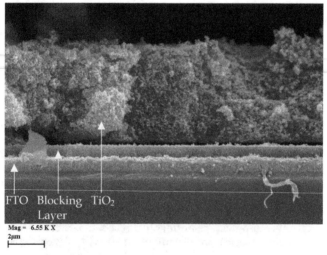

Fig. 3. Nanocrystalline TiO_2 film made of Degussa-P25 powder. A TiO_2 blocking layer is also present.

Indeed, the sol-gel method has led to the synthesis of a great variety of materials, the range of which is continuously expanding. Thus the simple incorporation of organic dopants as well as the formation of organic/inorganic nanocomposites offers the possibility of efficient dispersion of functional compounds in gels, it allows modification of the mechanical properties of the gels and provides materials with very interesting optical properties. A typical sol-gel route for making oxide matrices and thin films is followed by hydrolysis of alkoxides, for example, alkoxysilanes, alkoxytitanates, etc (Brinker & Scherer, 1990).

However, a review of the recent literature reveals an increasing interest in another sol-gel route based on organic acid solvolysis of alkoxides (Birnie & Bendzko, 1999; Wang et al., 2001). This second method seems to offer substantial advantages in several cases and it is becoming the method of choice in the synthesis of organic/inorganic nanocomposite gels. As it has been earlier found by Pope and Mackenzie (Pope & Mackenzie, 1986) and later verified by others, organic (for example, acetic or formic) acid solvolysis proceeds by a two step mechanism which involves intermediate ester formation (Ivanda et al., 1999). Simplified reaction schemes showing gel formation either by hydrolysis or organic acid solvolysis are presented by the following reactions. (Note that in these reactions only one metal-bound ligand is taken into account, while acetic acid (AcOH) is chosen to represent organic acids in organic acid solvolysis):

Hydrolysis	$\equiv M\text{-}OR + H_2O \rightarrow \equiv M\text{-}OH + ROH$	(7a)
Polycondensation	$\equiv M\text{-}(OH) \rightarrow \text{-}M\text{-}O\text{-}M\text{-} + H_2O$	(7b)
Acetic acid solvolysis	$\equiv M\text{-}OR + AcOH \rightarrow \equiv M\text{-}OAc + ROH$	(8a)
	$ROH + AcOH \rightarrow ROAc + H_2O$	(8b)
	$\equiv M\text{-}OAc + ROH \rightarrow ROAc + \equiv M\text{-}OH$	(8c)
	$\equiv M\text{-}OR + \equiv M\text{-}OAc \rightarrow ROAc + M\text{-}O\text{-}M$	(8d)

where M is a metal (for example, Si or Ti) and R is a short alkyl chain (for example, ethyl, butyl, or isopropyl). Hydrolysis (7a) produces highly reactive hydroxide species M-OH, which, by inorganic polymerization, produce oxide, i.e. M-O-M, which is the end product of the sol-gel process. More complicated is acetic acid solvolysis (8) where several different possibilities may define different intermediate routes to obtain oxide. Reaction (8a) is a prerequisite of the remaining three reactions. Occurrence of reaction (8b) would mean that water may be formed which may lead to hydrolysis. Reaction (8c) would create reactive M-OH which would form oxide, while reaction (8d) directly leads to oxide formation. The above possibilities have been demonstrated by various researchers by spectroscopic techniques. However, there still exists a lot of uncertainly and there is no concrete model to describe a well established procedure leading to oxide formation by organic acid solvolysis. For this reason, more work needs to be carried out on these systems. Reactions (8) reveal one certain fact. The quantity of acetic acid in solution will be crucial in affecting intermediate routes. Thus reaction (8b) is possible only if an excess of acetic acid is present. Also the quantity of acetic acid will define whether the solvolysis steps will simultaneously affect all available alkoxide ligands or will leave some of them intact and subject to hydrolysis reactions. Figure 4 shows a SEM cross sectional image of a mesoporous TiO_2 film prepared by sol-gel method with dip-coating. The average pore size is lower than 10 nm and particle diameter 10-12 nm. In this case, it is proved that no compact TiO_2 layer acting as "blocking layer" is needed for high performance DSSCs.

From previous paragraphs is obvious that nanoparticulate films are the common choice in photoelectrode preparation for use in DSSCs. However, the nanoparticulate films are not thought to be ideal in structure with regard to electron transport. For this reason, recent developments in nanostructured electrodes are proposed such as nanowires, nanotubes, nanorods which belong to 1-Dimensional structures in contrast to 3-D structures referred to films consisted of nanoparticles.

Fig. 4. TiO₂ nanocrystalline film made of sol-gel procedure (a) cross sectional image and (b) higher magnification of the film.

One-dimensional nanostructures might provide direct pathways for electron transport in DSSCs and ~25 µm thick film consisting of ZnO nanowires in diameter of ~130nm was mentioned to be able to achieve a surface area up to one-fifth as large as a nanoparticle film used in the conventional DSSCs (Law et al., 2005).

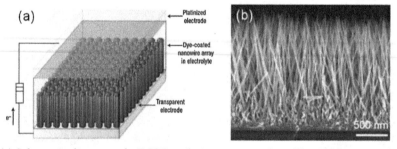

Fig. 5. (a) Schematic diagram of a DSSC with titania nanotubes, (b) a SEM image of titania nanotubes taken from reference (Zhang & Cao, 2011).

Moreover, the low manufacturing cost by using roll-to roll coating process creates the need of replacing the glass substrate with light weighted flexible plastic electrodes, expanding this way the area of DSSCs' applications. Flexible plastic electrodes like polyethylene terephthalate sheet coated with tin-doped indium oxide (PET-ITO) appear to possess many technological advantages (no size/shape limitations, low weight, high transmittance) as they present very low production cost in relation to F:SnO₂ (FTO) conductive glasses. The use of such plastic substrates requires that all processes needed for the fabrication of DSSC, including the formation of TiO₂ nanocrystalline films, to be designed at temperatures lower than 150⁰C. In the direction of replacing the glass substrates with flexible plastics, mesoporous TiO₂ films have to be prepared at low temperature and also with nanocrystalline dimensions for better efficiency to energy conversion. So far, the methods that obtain the most-efficient TiO₂ films for DSSCs have been based on high-temperature calcination. High-temperature annealing, usually at 450-500⁰C, is necessary to remove organic material needed to suppress agglomeration of TiO_2 particles and reduce stress during calcination for making crack-free films with good adhesion on substrates. Besides, high-temperature treatment of films promotes crystallinity of TiO_2 particles and their chemical interconnection for better electrical connection. Low sintering temperature yields titania nanocrystalline films with high active surface area but relatively small nanocrystals with many defects and poor interconnection, thus lower conductivity. High sintering temperature for TiO_2 films is then the most efficient method for the preparation of high performance DSSCs but it is also a cost intensive process. In addition, high temperature treatment of TiO_2 films cannot be applied to flexible plastic electrodes which in recent years emerge as an important technological quest. Different approaches appear in the literature to avoid high temperature annealing of thick and porous TiO_2 films. Among a variety of methods used for the low-temperature treatment of TiO_2 films like hydrothermal crystallization (Huang et al., 2006), chemical vapor deposition of titanium alkoxides (Murakami et al., 2004), microwave irradiation (Uchida et al., 2004), ultraviolet light irradiation treatment (Lewis et al., 2006), and sol-gel method (Stathatos et al., 2007), the efficiency of DSSCs employing ITO-PET substrates was in the range of 2-3% at standard conditions of 100 mW/cm^2 light intensities at AM 1.5. A very simple and also benign method for the formation of pure TiO_2 nanoparticles surfactant-free films of nanocrystalline TiO_2 at room temperature with excellent mechanical stability is the mixture of a small amount of titanium isopropoxide with commercially available P25-TiO_2 (surface area of 55 m^2/g, mean average particle size of 25 nm and 30/70% rutile/anatase crystallinity) powder. The hydrolysis of the alkoxide after its addition helps to the chemical connection between titania particles and their stable adhesion on plastic or glass substrate without sacrificing the desired electrical and mechanical properties of the film. Promising results have obtained by the use of this method.

3.2 Sensitizers

The dye plays the important role of sensitizing the semiconductor in the visible and infrared region of solar light. For this reason several requirements have to be succoured at the same time such as, broad absorption spectrum, good stability, no toxicity, good matching of the HOMO, LUMO levels of the dye with semiconductor's bottom edge of conduction band and chemical potential of redox system of the electrolyte. Besides, the chemical bonding between the dye and semiconductor's surface is absolutely necessary for effective electron transfer. The ideal sensitizer for nanocrystalline TiO_2 particles has to absorb all the light below a threshold wavelength of about 900nm. Moreover it has to carry out carbolxylate or phosphonate groups which are permanently grafted on oxide surface by chemical bonds so as after excitation to inject electrons into the semiconductor with a quantum yield close to unity. The stability of the sensitizer is ensured by 100 million turnover cycles which refer to approximate twenty years of light soaking (Grätzel & Durrant, 2008). The common sensitizers for DSSCs are ruthenium complexes with bipyridine ligands and they follow the structure $ML_2(X)_2$ where L is the organic ligand and M the metal ion (either Ru or Os) and X can be cyanide, thiocarbamate or thiocyanate groups. Electron transfer from sensitizer to semiconductor after optical excitation is based on metal to ligand charge transfer and then the transfer to the semiconductor via the chemical bond between them. The N3 dye (*cis*-bis(isothiocyanato)bis(2,2'-bipyridyl-4,4'-dicarboxylato)-ruthenium(II)) was first reported as the most efficient sensitizer for DSSCs (Nazeeruddin et al., 1999). Then Black Dye [*cis*-

diisothiocyanato-bis(2,2'-bipyridyl-4,4'-dicarboxylato)   ruthenium(II)   bis(tetrabutyl
ammonium) was also introduced by Grätzel and co-workers as a most efficient sensitizer
because it covers solar light in longer wavelengths than N3 (Nazeeruddin et al., 2001).
Modified N3 with tetrabutyl ammonium groups (N719) triisothiocyanato-(2,2':6',6"-
terpyridyl-4,4',4"-tricarboxylato) ruthenium(II) tris(tetra-butyl ammonium) was finally
found to be the most applicable dye in DSSCs' technology as it enhances the open circuit
voltage of the cells of at least 15%. Next generation of dyes is based on the formula of N3
while it contains different size groups on the ligands covering two basic demands: (a)
chemical stability and good penetration of electrolyte because of suitable organic groups (b)
absorbance in longer wavelengths. Recent years the combination of dye properties with
organic p-type semiconducting side groups seems to attract much attention. Another case of
sensitizers is pure organic dyes in replacement of costly ruthenium complexes. Metal free
sensitizers for DSSCs are referred: hemicyanines, indoline dyes, phthalocyanines,
coumarins, perylene derivatives etc. Promising results have been obtained where in the case
of D149 indoline dye an efficiency of 9.5% was recorded while SQ2 (5-carboxy-2-[[3-[(2,3-

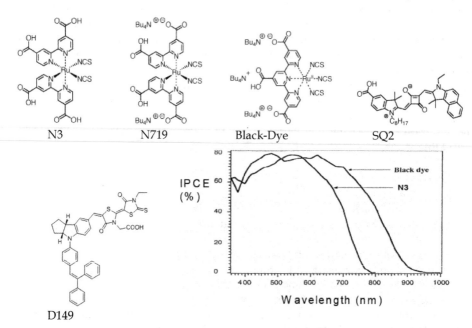

N3   N719   Black-Dye   SQ2

D149

Fig. 6. Incident Photon to current efficiency for N3 and Black dye (Grätzel, 2006).

dihydro-1,1-dimethyl-3-ethyl-1H-benzo[e]indol-2-ylidene)   methyl   ]-2-hydroxy-4-oxo-2-
cyclobuten-1-ylidene]methyl]-3,3-dimethyl-1-octyl-3H-indolium) an efficiency of 8% was
also recorded (Goncalves et al., 2008).   Finally the strengths and weaknesses of organic
dyes in DSSCs are the followings:
The strengths are:
• They exhibit high absorption coefficient (abundant π→π* within molecules)
• it is easy to design dyes with various structures and adjust absorption wavelength
  range

- uses no metal and they have no limitation to resources
- lower cost than organic metal dye and enables synthesis.

While the weaknesses of organic dyes are:

- still lower efficiency than organic metal dye
- short post-absorption excited state (π^*) lifetime
- narrow absorption spectrum wavelength in visible ray field that it is difficult to absorb light from all visible range.

3.3 Electrolytes

The key composition elements for DSSC include fluorinated tin oxide (FTO) which is used for either electrode substrate, nanoparticulated oxide semiconductor layer like TiO_2 and ZnO, sensitizer, metallic catalysts like platinum which plays the role of the opposite electrode and the electrolyte which includes redox couple and it is positioned between the two electrodes. The composition and the form of the electrolyte have great affect on the total energy conversion efficiency. The majority of the proposed DSSCs is based on liquid electrolytes with a variety of solvents where an overall maximum efficiency of ~12% was finally achieved. Nevertheless, there are still questions which own an answer about the stability and sealing in order to prevent the leakage of the solvent. Solid or quasi solid electrolytes could be an answer to the questions. In the case of solid electrolytes we may refer polymeric materials which incorporate the iodide/triiodide redox, organic hole transporting materials or inorganic p-type semiconductors. As quasi-solid electrolytes we can refer composite organic/inorganic materials which appear as a gel or highly viscous ionic liquids.

3.3.1 Liquid electrolytes

The electrolyte is generally composed with oxidation-reduction of I^-/I_3^- where LiI, NaI, alkyl ammonium iodine or imidazolium iodine is used for materials of I^- ion. For instance, 0.1M LiI, 0.05M I_2, and 0.5M *tert*-butyl pyridine (TBP) are mixed in acetonitrile solution or 3-methoxypropionitrile, propylenecarbonate, γ-butyrolaqctone, N-methylpyrrolidone as alternative solvents. I^- ion is responsible for offering electrons for holes generated in dye molecule's HOMO level, whereas the oxidized I_3^- ion accepts electrons that reach counter electrode to be reduced (Snaith & Schmidt-Mende, 2007).

3.3.2 Solid electrolytes

When it is referred the use of solid electrolytes it is generally accepted that people mean the use of p-type semiconducting materials either organic or inorganic. In the case of organic materials the most popular is spiro-MeOTAD which initially proposed by Grätzel and co-workers (Kruger et al., 2002). The recorded efficiency was about 4% while new organic semiconductors were appeared in the meanwhile. Polymer based solid electrolytes are usually referred as efficient alternatives to liquid based electrolytes but the efficiency is still poor. Polymer usually containing polyether units can be used as solid electrolytes in DSSCs. These types of electrolytes are solid ionic conductors prepared by the dissolution of salts in a suitable high molar mass polymer containing polyether units (de Freitas et al., 2009).

In the case of inorganic materials, CuSCN (O'Regan & Schwartz, 1998) and CuI (Tennakone et al. 1995) are the most popular, but the efficiency of the solar cells is lower than 2% because of the poor contact between TiO_2 and p-type inorganic semiconductor. Optimized interface between the two different types of semiconductors and instability problems of copper based p-type semiconductors have to be improved before the p-n junction between them to be more efficient.

3.3.3 Quasi-solid electrolytes

In some cases the very viscous ionic liquids but in most cases the composite organic/inorganic materials are referred as quasi-solid electrolytes. Nanocomposite organic/inorganic materials are constituted of two interpenetrating subphases which are mixed in nanoscale. The organic subphase is usually consisted of few surfactants or polyether chains and the inorganic subphase is made of an inorganic network which typically is silicon dioxide or titanium dioxide. Such nanocomposite gels can accommodate appropriate solvents within the organic subphase (that is within the pores left by the inorganic network) so that ionic conductivity can be raised to a satisfactory degree. The design and synthesis of such materials makes for fascinating research with numerous scientific and technological implications in iono-electronics, mechanics and optics. There are two prospects of making organic/inorganic blends which are depended on the specific interactions between the two subphases (scheme 1). Such blends which were obtained by simply mixing of the two components together, characterized as materials of *Class I* corresponding no covalent or iono-covalent bonds. In these materials the various components only exchange weak interactions such as hydrogen bonding Van Der Waals interactions or electrostatic forces. On the other hand, materials which are formed by chemically bonding between the two subphases are characterized as *Class II* (hybrid materials). *Class II* materials organic/inorganic components are linked through strong chemical bonds e.g. covalent, iono-covalent or Lewis acid-base bonds. Usually materials of *Class II* have better mechanical properties than *Class I* as they present rubbery behavior (Stathatos, 2005).

Sol-Gel chemistry allows the combination at the nanosize level of inorganic and organic since solubility of most organic substances, especially, hydrophobic ones, is limited in pure oxides (e.g. SiO_2) causing migration and aggregation with subsequent decrease of their functionality. Nanocomposite gels made of the two different subphases the oxide network, as the inorganic subphase and the polymer or surfactant as the organic subphase mixed in nanoscale providing in this way an access to an immense new area of materials science. In principal, in sol-gel chemistry a metal alkoxide (including silicon alkoxides) is hydrolyzed and the subsequent inorganic polymerization leads to the formation of the corresponding oxide with an oncoming condensation of the material. The whole process is carried out at ambient conditions. The process can be summarized to reactions 7 mentioned before.

An alternative route to the oxide synthesis is the slow water release in the solution with no initial water addition into the solution. In this case the existence of an organic acid in the sol is indispensable, typically acetic or formic acid as initially referred in reactions 8.

(In reactions 7a and b, as well as in the above reaction 8, only one of usually four reacting alkoxy groups is taken into account, for reasons of simplicity). Reactions 8 show that the end product of the sol-gel process is -M-O-M-, which can be obtained by successive chemical reactions. Metal ester (M-OAc) as a result of (2) can react with the metal alkoxide forming -M-O-M-, additionally; water released through esterification reaction 3 can yield oxide by the hydrolysis route. When ethanol is introduced in the sol, which is a common recipe in many works, even more water can be released by direct esterification reaction between the ethanol (EtOH) and the acetic acid (AcOH). Furthermore, intermediate M-OAc ester or -M-O-Moligomers may create entities which offer polymorphism to the sol-gel evolution. Thus the presence of a self-organizing agent, e.g. a surfactant, plays a crucial role in organizing the structure of the material and in creating well defined and reproducible nanophases. Slow

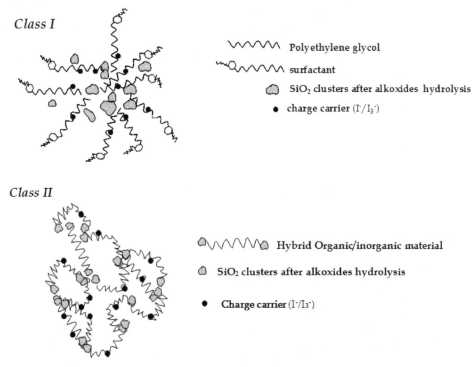

Scheme 1. (a) Class I of composite organic-inorganic electrolyte (b) Class II of hybrid organic-inorganic electrolyte

water release, organic acid solvolysis and surfactant organization are then the key factors that dictate the structure and the quality of the nanocomposite organic/inorganic gel. A different approach to the gel process of quasi-solid electrolytes is the use of modified materials (Jovanovski et al., 2006) with silicon alkoxy-groups which may easily hydrolyzed and finally lead to a gel formation (e.g. scheme 2). The modified materials could be a series of additives usually employed in liquid electrolytes e.g. benzyl-imidazoles for open circuit voltage enhancement which are now bearing alkoxy-groups for jellifying process.

As a consequence, gel electrolytes are roughly distinguished into three categories: (1) One way to make a gel electrolyte is to add organic or inorganic (or both) thickeners. Such materials may be long-chain polymers like poly (ethylene oxide) or inorganic nanoparticles like titania or silica; (2) A second way is to introduce a polymerizable precursor into the electrolyte solution and polymerize the mixture in situ; (3) a third route is to produce a gel incorporating the I^-/I_3^- redox couple through the sol-gel process by using a sol-gel precursor, like a titanium or silicium alkoxide. This precursor may be a functionalized derivative of one of the components of the electrolyte. This last method has been very successful since the sol-gel process leads to the formation of nanocomposite organic-inorganic materials. Such materials are composed of an inorganic subphase, which binds and holds the two electrodes together and seals cell and an organic subphase, which assures dispersion of ionic species and supports ionic conductivity. The whole composition is compatible with titania nanocrystalline electrode and provides good electrical conduct and finally satisfactory ionic conductivity. Such cells are easy to make. After dye-adsorption on titania electrode, it suffices to place a small drop of the sol on the surface of the electrode and then press the counter electrode on the top by hand under ambient conditions. The two electrodes bind together while the fluid sol enters into titania nanoporous structure and achieves extensive electrical conduct.

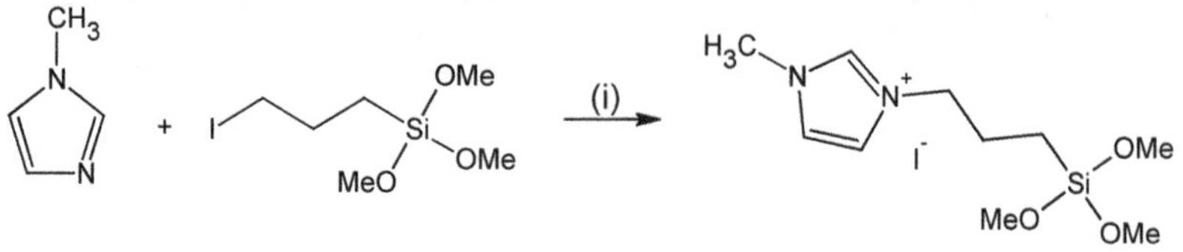

Scheme 2. Example of a hybrid organic/inorganic material used as quasi-solid electrolyte. (Jovanovski et al., 2006)

3.4 Positive electrode (catalysts)

The counter electrode is one of the most important components in the dye-sensitized solar cell. The major role of the counter electrode in addition to the cell finalization is the reduction of the redox species used as a mediator in regenerating the sensitizer after electron injection, or collection of the holes from the hole conducting material in a solid-state DSSC. Counter electrodes of dye-sensitized solar cells can be prepared with different materials and methods. Platinum, graphite, activated carbon, carbon black, single-wall carbon nanotubes, poly(3,4-ethylenedioxythiophene) (PEDOT), polypyrrole, and polyaniline can be used as catalysts for the reduction of triiodide. Moreover, for the ultimate in low-cost counter electrodes, it is also referred in literature a carbon-black-loaded stainless steel electrode for use as a novel counter electrode (Murakami & Grätzel, 2008).

4. Manufacturing of Dye sensitized solar cells

While many research groups investigate the working principles of DSSCs and new developments have been achieved concerning their efficiency and large scale applications, new companies founded in the meanwhile try to carry DSSC technology in market place evaluating all process steps are needed for industrial production. Experimental results for small size solar cells cannot directly applied in large scale DSSCs as the efficiencies measured for small size solar cells cannot be repeated in large scale (Späth et al., 2003). This is caused by the high internal resistance of FTO substrates which eliminates their efficiency and it is found to be drastically decreased. Below are referred some of the main issues which have to be taken into account before DSSCs go to a production line:

- Large area deposition of TiO_2 layers. The layers have to be homogeneous and uniform
- New methods for dye staining and electrolyte filling
- Electrical interconnection of individual cells. A major factor for limited efficiency of the DSSC is the ineffective contacts on FTO glass. The external connections of the individual cells are also a problem.

- Sealing process for modules in case of liquid electrolytes
- Long-term stability of at least 10 years for outdoor use.
- Evaluation costs, which is believed to be approximately 10% of that needed for silicon solar cells

Moreover, for the durability of the cells, it was considered necessary to replace the liquid electrolyte with quasi-solid state, solid, polymer electrolytes or p-type inorganic semiconductors as the manufacturing cost and convenience of preparation is highly simplified. However, the low manufacturing cost by using the recently developed roll-to-roll coating process especially for DSSCs creates the need of replacing the glass substrate with light weighted flexible plastic electrodes, expanding this way the area of DSSCs' applications. Three basic structures for large scale DSSCs are proposed either for high current (cells in parallel) or for high voltage (cells in series) collection:

4.1 Monolith module

The monolith modules have similar structure to amorphous silicon modules that are currently used in the market. Monolith modules use a single sheet of conductive glass (FTO) where successive layers of TiO$_2$ are formed on the substrate. Prior films deposition the glass is scribed by a laser in order to isolate one cell from the other. Direct wiring is not needed in this type of modules. It records relatively broad available area and high conversion efficiency, and enables simultaneous production of multiple cells (Wang et al., 2010). Usually, the TiO$_2$ stripes are formed according to screen printing method. It also entails shortcomings that it must secure even efficiency of each cell since it is in serial connection mode, it may by damaged due to relatively weak surface, and it rather has low transmittance. The figure 7 appears below, illustrates the manufacturing method of a monolith module:

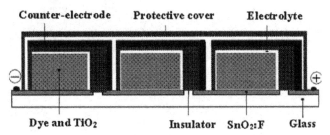

Fig. 7. Monolithic DSSC module, taken from reference (Dai et al., 2008)

4.2 Z-module

It concerns series connections of individual cells and consists of two opposing electrodes with inner-connections between neighbouring cells by a metal conductor. A sealing material is needed to protect the metal conductor from corrosion by iodide ions. It has wide available area to realize relatively high photoelectric conversion efficiency and enables transparent and double sided cell production. The Z-module entails shortcomings that it is difficult to match the junction for large-area cell production, it needs to reduce each cell efficiency deviation due to the series connection, and it is greatly affected by inner-connector reliability and conduction property (Sastrawan et al., 2006). The advantage of Z-type module fabrication is the high voltage output. On the other hand, the disadvantage of this connection is the low active area and overall efficiency because of the complicated structure and the resulting high series resistance. Figure 8b illustrates a possible Z-module manufacturing process.

4.3 W-module

It also concerns a series connection of the cells while inner-connections are avoided. Unlike the z-module and monolith module that are manufactured in the same direction to the adjacent cell, the W-module secures a structure that are in an opposite direction of the adjacent cell. Hence, it does not require direct wiring or bus electrode and realizes high reliability since the inner-cell contact occurs directly on the substrate. It also secures a structure to maximize the available area to record relatively high conversion efficiency. It is able to simultaneously manufacture multiple cells and can also manufacture cell and module at the same time. It entails shortcomings as well that it needs to adjust the output since the amount of light absorption is different in each serially connected cell and the colorants of adjacent cells are different. The main disadvantage of this structure is the differences referred for currents of single cells which are illuminated from back side and front size resulting different values of current because of the different light transmittance of the two cells.

4.4 Parallel-module

A simple fabrication procedure of making DSSC modules is parallel connection. According to this structure small cells in the form of long stripes are connected in parallel.

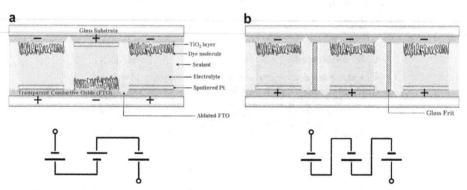

Fig. 8. (a) W-Module (b) Z-Module DSSC (Seo et al., 2009)

In this structure parallel grids utilizing conductive fingers to collect current are printed on the two electrodes of the cells. The printing method of current collectors is quite similar to that applied in conventional photovoltaics based on silicon. Common metals used as current collectors reducing the distance of electron transfer and internal resistance of FTO glass are: Ag, Cu, Ni, Ti. The solar cell efficiency is considerably reduced when it is converted to module despite the high unit cell efficiency. This is because of the increased possibility to lose electrons, which are created by light absorption, through either internal defect or recombination with hole at interface with other materials during the delivery when the electrode area to absorb light enlarges. Therefore, the efficiency radically decreases when DSSC active electrode width becomes greater than 1 cm (Wang et al., 2010).

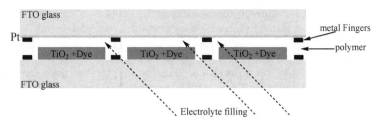

Fig. 9. Parallel connection of DSSC in a module

It is essential to design and manufacture effective packaging system along with designing photovoltaic absorption-use electrode and charge collection-use grid in order to allow the electron flow to collect without losses in large-areas like module. Mainly opposed cell module has been manufactured and researched since 1995 until now. The opposed cell module used ceramic fragment paste (glaze) or polymer in order to protect the conductive internal pattern from electrolytes. Such opposed cell R&D activities slowed down until 2001 and newly begun afterwards (Displaybank, 2010). The parallel type module records broad active area and high conversion efficiency. Large-area photoelectric chemical solar cell must use transparent electrode which has weaker electric conductivity than the metal wiring that it requires a grid to play a role of charge collection to realize smooth electron delivery. Therefore, the large-area solar cell exhibits different carrier generation and delivery from the unit cell.

Fig. 10. (a) artistic DSSC module by Sony (http://www.sony.net/SonyInfo/technology/technology/theme/solar_01.html) and (b - c)DSSC panels made of Fujikura (http://www.fujikura.co.jp/eng/rd/field/mt.html)

A general grid in DSSCs mainly uses metal material. This is connected to active area decrease and becomes the factor to increase the generation unit cost of cell. Therefore, it is essential to secure effective module design and manufacturing technology for commercialization.

The DSSC is manufactured by a process that is relatively simpler than the conventional solar cells made of silicon and compound semiconductor solar cells, but it entails a shortcoming to generate metal corrosion when using the metal with outstanding electric conductivity as grid due to iodine based electrolyte. Therefore, the DSSC is in need for electrolyte development with outstanding activity without corrosive property or metal development with outstanding electric conductivity without being separated or corroded from the electrolyte. The inverter development must progress to be appropriate for DSSC which secures electric property that is different from the conventional silicon based solar cell.

In order to accomplish this, a circuit must be composed to match arrangement and response properties of DSSC module. The inverter technology development maximizes the DSSC power generation efficiency. The system is matched to the solar cell's generation property in order to effectively supply the power of electric condenser, which stores electricity generated during daytime, at desirable time. The commercialization of DSSC requires power system development together with unit cell efficiency enhancement technology development. The DSSC commercialization is delayed due to unprepared peripheral technologies despite the fact that its current power generation unit cost can realize commercialization (Displaybank, 2010). Right now, the ultra small high efficiency inverter technology is insufficient for module/system efficiency enhancement and manufacturing technology of module which can be installed in targets such as buildings.

5. Outlook

The quest and demand for clean and economical energy sources have increased interest in the development of solar applications. DSSCs have proved to be an alternative approach to the conventional silicon based solar cells. Research on DSSCs has grown rapidly in the recent years due to the several attractive figures of this interesting field: The international awareness of the necessity to develop new technologies in Renewable Energy Resources; The need of easy and inexpensive procedures for fabricating Solar cells; The fact that DSSCs can be transparent so that they can be integrated into mobile or immobile constructions as Photovoltaic Windows. All these features are carried by DSSCs and for this reason they are popular and they are expected to be even more popular in the near future. Till then some issues have to be overcome in order this technology considerably has commercial interest. The overall efficiency of ~12% for small size cells (~0.2cm^2) which substantially gets lower (~5%) when modules of DSSCs are prepared is a reason for further improvement is many stages of cells preparation. Improvement of the cells is focused on (a) the enhancement of electron transport and electron lifetime in the mesoporous metal oxide (b) design of new high-extinction coefficient dyes which will effectively cover the whole range of visible light and near infra red and finally (c) new stable solid electrolytes which will have effective penetration into semiconductors pores and enhanced. Moreover, in large scale applications the effective collection of the current is also an issue as the internal resistance from conductive glass substrates and metal grids which are necessary for current collection and need to be covered from corrosive electrolyte, drastically eliminate the DSSCs efficiency if they are not seriously taken into account.

6. References

Barbé, C.J.; Arendse, F.; Comte, P.; Jirousek, M.; Lenzmann, F.; Shklover, V.; Grätzel, M. (1997). Nanocrystalline titanium oxide electrodes for photovoltaic applications. *Journal of the American Ceramic Society* vol. 80, pp. 3157-3171

Birnie. DP.; Bendzko, NJ. (1999). H-1 and C-13 NMR observation of the reaction of acetic acid with titanium isopropoxide. *MATERIALS CHEMISTRY AND PHYSICS* Vol. 59, pp. 26-35

Bisquert, J.; Cahen, D.; Hodes, G.; Ruhle, S.; Zaban, A. (2004). Physical Chemical Principles of photovoltaic conversion with nanoparticulate, mesoporous dye-sensitized solar cells. *J. Phys. Chem.* vol. 108, pp. 8106-8118

Brinker, C.J.; Scherer, G.W. (1990). Sol-Gel Science, The physics and Chemistry of Sol-Gel Processing. *Academic Press, Inc*

Dai, S.; Weng, J.; Sui, Y.; Chen, S.; Xiao, S.; Huang, Y.; Kong, F.; Pan, X.; Hu, L.; Zhang, C.; Wang K. (2008). The design and outdoor application of dye-sensitized solar cells. *Inorganica Chimica Acta.* Vol. 361, pp. 786-791

de Freitas, J.N.; Nogueira, A.F.; De Paoli, M.A. (2009). New insights into dye-sensitized solar cells with polymer electrolytes. *Journal of Materials Chemistry.* Vol. 19, pp. 5279-5294

Displaybank (2010). DSSC Technology Trend and Market Forecast 2009-2013 (*www.displaybank.com*)

Hagfeldt, A.; Grätzel, M. (2000). Molecular photovoltaics. *Acc. Chem. Res.* Vol. 33, pp. 269-277

Jose, R.; Thavasi, V.; Ramakrishna S. (2009). Metal Oxides for Dye-Sensitized Solar Cells. *J. Am. Ceram. Soc.* vol. 92, pp. 289-301

Goncalves, L.M.; de Zea Bermudez, V.; Ribeiro, H.A.; Mendes, A.M. (2008). Dye-sensitized solar cells: A safe bet for the future. *Energy Environ. Sci.* vol. 1, pp. 655-667

Goldstein, J.; Yakupov, I.; Breen, B. (2010). Development of large area photovoltaic dye cells at 3GSolar. *Solar Energy Materials and Solar Cells.* Vol. 94, pp. 638-641

Grätzel, M. (2006). Photovoltaic performance and long-term stability of dye-sensitized mesoscopic solar cells. *C.R. Chimie*, vol.9, pp.578-583

Grätzel, M.; Durrant, J.R. (2008). Dye sensitized mesoscopic solar cells. *Series on Photoconversion of Solar Energy Vol.3 Nanostructured and Photoelectrochemical systems for solar photon conversion.* Imperial College Press.

Hinsch, A.; Brandt, H.; Veurman, W.; Hemming, S.; Nittel, M.; Wurfel, U.; Putyra, P.; Lang-Koetz, C.; Stabe, M.; Beucker, S.; Fichter, K. (2009). Dye solar modules for façade applications: Recent results from project ColorSol. *Solar Energy Materials and Solar Cells.* Vol. 93, pp. 820-824

Huang, C.-Y.; Hsu, Y.-C.; Chen, J.-G.; Suryanarayanan, V.; Lee, K.-M.; Ho, K.-C. (2006). The effects of hydrothermal temperature and thickness of TiO_2 film on the performance of a dye-sensitized solar cell. *Solar Energy Materials and Solar Cells.* Vol. 90, p. 2391

Ivanda, M.; Musić, S.; Popović, S.; Gotić, M. (1999). XRD, Raman and FT-IR spectroscopic observations of nanosized TiO_2 synthesized by the sol–gel method based on an esterification reaction. *Journal of Molecular Structure.* Vol. 480-481, pp. 645-649

Jovanovski, V.; Stathatos, E.; Orel, B.; Lianos, P. (2006). Dye-sensitized solar cells with electrolyte based on a trimethoxysilane-derivatized ionic liquid. *Thin Solid Films.* Vol. 511, pp. 634-637

Kruger, J.; Plass, R.; Gratzel, M.; Matthieu, H. (2002). Improvement of the photovoltaic performance of solid-state dye-sensitized device by silver complexation of the sensitizer cis-bis(4,4 '-dicarboxy-2,2 ' bipyridine)-bis(isothiocyanato) ruthenium(II). *Applied Physics Letters.* Vol. 81, pp. 367-369

Law, M.; Greene, LE.; Johnson, JC.; Saykally, R.; Yang, PD. (2005). Nanowire dye-sensitized solar cells. *Nature Materials.* Vol. 4 pp. 455-459

Murakami, TN.; Gratzel, M. (2008). Counter electrodes for DSC: Application of functional materials as catalysts. Inorganica Chimica Acta. Vol. 361, pp. 572-580

Murakami, T.N.; Kijitori, Y.; Kawashima N.; Miyasaka, T. (2004). Low temperature preparation of mesoporous TiO2 films for efficient dye-sensitized photoelectrode by chemical vapor deposition combined with UV light irradiation. *J. Photochem. Photobiol. A: Chem.* Vol. 164, p. 187

Nazeeruddin, M. K.; Kay, A.; Rodicio, I.; Humphry-baker, R.; Muller, E.; Liska, P. (1993). Conversion of light to electricity by cis-X2-bis(2,2'-bipyridyl-4,4'-dicarboxylate)ruthenium(II) charge-transfer sensitizers (X = Cl-, Br-, I-, CN- and SCN-) on nanocrystalline TiO2 electrodes. *J. Am. Chem. Soc.* Vol. 115, pp.6382-6390

Nazeeruddin, M. K.; Pechy, P.; Renouard, T.; Zakeeruddin, S. M.; Humphry-Baker ,R.; Comte, P. (2001). Engineering of efficient panchromatic sensitizers for nanocrystalline TiO2-based solar cells. *J. Am. Chem. Soc.* Vol. 123, pp. 1613-1624.

Papageorgiou, N.; Athanassov, Y.; Armand, M.; Banhote, P.; Lewis, L.N.; Spivack, J.L.; Gasaway, S.; Williams, E.D.; Gui, J.Y.; Manivannan, V.; Siclovan, O.P. (2006). A novel UV-mediated low-temperature sintering of TiO2 for dye-sensitized solar cells. *Solar Energy Materials and Solar Cells.* Vol. 90 p. 1041

O'Regan, B.; Grätzel, M. (1991). A low-cost, high-efficiency solar-cell based on dye-sensitized colloidal TiO2 films. *Nature,* vol. 353, pp.737-740

O'Regan, B.; Schwartz, DT. (1998). Large enhancement in photocurrent efficiency caused by UV illumination of the dye-sensitized heterojunction TiO2/RuLL ' NCS/CuSCN: Initiation and potential mechanisms. *Chemistry of Materials.* Vol. 10, pp. 1501-1509

Peter, L.M.; Wijayantha, K.G.U. (2000). Electron transport and back reaction in dye sensitised nanocrystalline photovoltaic cells. *Electrochim. Acta.* vol. 45, pp. 4543-4551

Petterson, H.; Azam, A.; Grätzel, M. (1996). The Performance and Stability of Ambient Temperature Molten Salts for Solar Cell Applications. *Electrochem. Soc.* Vol. 143, pp. 3099-3108

Pope, E.J.A.; Mackenzie, J.D. (1986). Sol-gel processing of silica: II. The role of the catalyst. *Journal of Non-Crystalline Solids.* Vol. 87, pp. 185-198

Sastrawan, R.; Beier, J.; Belledin, U.; Hemming, S.; Hinsch, A.; Kern, R.; Vetter, C.; Petrat, F.M.; Prodi-Schwab, A.; Lechner, P.; Hoffmann, W. (2006). A glass frit-sealed dye solar cell module with integrated series connections. *Solar Energy Materials and Solar Cells.* Vol. 90 pp. 1680-1691

Sayama, K.; Sugihara, H.; Arakawa, H. (1998). Photoelectrochemical properties of a porous Nb$_2$O$_5$ electrode sensitized by a ruthenium dye. *Chem. of Materials* vol. 10, pp. 3825–3832.

Seo, H.; Son, M.; Hong, J.; Lee, D.-Y.; An, T.-P.; Kim, H.; Kim, H.-J. (2009). The fabrication of efficiency-improved W-series interconnect type of module by balancing the performance of single cells. *Solar Energy.* Vol. 83, pp. 2217-2222.

Snaith, H.J.; Schmidt-Mende, L. (2007). Advances in Liquid-Electrolyte and solid-state dye-sensitized solar cells. *Advanced Materials.* Vol. 19, pp. 3187-3200

Späth, M.; Sommeling, P.M.; van Roosmalen, J.A.M.; Smit, H.J.P.; van der Burg, N.P.G.; Mahieu, D.R.; Bakker, N.J.; Kroon, J.M. (2003). Reproducible manufacturing of Dye-sensitized solar cells on a semi-automated baseline. *Progress in Photovoltaics: Research and applications.* Vol. 11 pp. 207-220

Stathatos, E.; Lianos, P.; Tsakiroglou, C. (2004). Highly efficient nanocrystalline titania films made from organic/inorganic nanocomposite gels. *Microporous and Mesoporous Materials.* vol. 75, pp. 255-260

Stathatos, E.; Choi, H.; Dionysiou, D.D. (2007). Simple procedure of making room temperature mesoporous TiO$_2$ films with high purity and enhanced photocatalytic activity. *Environmental Engineering Science.* vol. 24, p. 13

Stathatos, E.; Lianos, P.; Del Monte, F.; Levy, D.; Tsiourvas, D. (1997). Formation of TiO$_2$ nanoparticles in reverse micelles and their deposition as thin films on glass substrates. *Langmuir.* Vol. 13, pp. 4295-4300

Stathatos, E. (2005). Organic-inorganic nanocomposite materials prepared by the sol-gel route as new ionic conductors in quasi solid state electrolytes.   Ionics. Vol. 11 pp. 140-145

Tennakone, K.; Kumara, GRRA.; Kottegoda, IRM.; Perera, V. P. S. (1999). An efficient dye-sensitized photoelectrochemical solar cell made from oxides of tin and zinc. Chemical Communications pp. 15-16

Tennakone, K.; Kumara, GRRA.; Kumarasinghe, AR.; Wijayantha, KGU.; Sirimanne, PM. (1995). A dye sensitized nano-porous solid state photovoltaic cell. *Semiconductor science and Technology.* Vol. 10 pp. 1689-1693

Uchida, S.; Tomiha, M.; Takizawa, H.; Kawaraya, M. (2004). Flexible dye-sensitized solar cells by 28 GHz microwave irradiation. *J. Photochem. Photobiol. A: Chem.* Vol. 164, p. 93

Wang, C.; Deng, ZX.; Li, YD. (2001). The synthesis of nanocrystalline anatase and rutile titania in mixed organic media. *Inorganic Chemistry.* Vol. 40, pp. 5210-5214

Wang, L.; Fang, X.; Zhang, Z. (2010). Design methods for large scale dye-sensitized solar modules and the progress of stability research. *Renewable and Sustainable energy reviews.* Vol. 14, pp. 3178-3184

Wu, J.; Lan, Z.; Hao, S.; Li, P.; Lin, J.; Huang, M.; Fang, L.; Huang, Y. (2008). Progress on the electrolytes for dye-sensitized solar cells. *Pure Applied Chemistry,* vol.80, No.11, pp.2241-2258

Zhang, Q.; Cao, G. (2011). Nanostructured photoelectrodes for dye-sensitized solar cells. *Nano Today* vol. 6, pp. 91–109

Zhang, W.; Cheng, Y.; Yin, X.; Liu, B. (2011). Solid-state dye-sensitized solar cells with conjugated polymers as hole-transporting materials. *Macromolecular Chemistry and Physics.* Vol. 212, pp. 15-23

Zhu, K.; Schiff, E.A.; Park, N.-G.; van de Lagemaat, J.; J. Frank, A. (2002). Determining the locus for photocarrier recombination in dye-sensitized solar cells. *Applied Physics Letters.* Vol.80, pp. 685-687

Xia, J.; Yanagida, S. (2011) Strategy to improve the performance of dye-sensitized solar cells: Interface engineering principle. *Solar Energy* in press available in www.sciencedirect.com

Permissions

The contributors of this book come from diverse backgrounds, making this book a truly international effort. This book will bring forth new frontiers with its revolutionizing research information and detailed analysis of the nascent developments around the world.

We would like to thank Garry Einicke, for lending his expertise to make the book truly unique. He has played a crucial role in the development of this book. Without his invaluable contribution this book wouldn't have been possible. He has made vital efforts to compile up to date information on the varied aspects of this subject to make this book a valuable addition to the collection of many professionals and students.

This book was conceptualized with the vision of imparting up-to-date information and advanced data in this field. To ensure the same, a matchless editorial board was set up. Every individual on the board went through rigorous rounds of assessment to prove their worth. After which they invested a large part of their time researching and compiling the most relevant data for our readers. Conferences and sessions were held from time to time between the editorial board and the contributing authors to present the data in the most comprehensible form. The editorial team has worked tirelessly to provide valuable and valid information to help people across the globe.

Every chapter published in this book has been scrutinized by our experts. Their significance has been extensively debated. The topics covered herein carry significant findings which will fuel the growth of the discipline. They may even be implemented as practical applications or may be referred to as a beginning point for another development. Chapters in this book were first published by InTech; hereby published with permission under the Creative Commons Attribution License or equivalent.

The editorial board has been involved in producing this book since its inception. They have spent rigorous hours researching and exploring the diverse topics which have resulted in the successful publishing of this book. They have passed on their knowledge of decades through this book. To expedite this challenging task, the publisher supported the team at every step. A small team of assistant editors was also appointed to further simplify the editing procedure and attain best results for the readers.

Our editorial team has been hand-picked from every corner of the world. Their multi-ethnicity adds dynamic inputs to the discussions which result in innovative outcomes. These outcomes are then further discussed with the researchers and contributors who give their valuable feedback and opinion regarding the same. The feedback is then collaborated with the researches and they are edited in a comprehensive manner to aid the understanding of the subject.

Apart from the editorial board, the designing team has also invested a significant amount of their time in understanding the subject and creating the most relevant covers. They scrutinized every image to scout for the most suitable representation of the subject and create an appropriate cover for the book.

The publishing team has been involved in this book since its early stages. They were actively engaged in every process, be it collecting the data, connecting with the contributors or procuring relevant information. The team has been an ardent support to the editorial, designing and production team. Their endless efforts to recruit the best for this project, has resulted in the accomplishment of this book. They are a veteran in the field of academics and their pool of knowledge is as vast as their experience in printing. Their expertise and guidance has proved useful at every step. Their uncompromising quality standards have made this book an exceptional effort. Their encouragement from time to time has been an inspiration for everyone.

The publisher and the editorial board hope that this book will prove to be a valuable piece of knowledge for researchers, students, practitioners and scholars across the globe.

List of Contributors

Ho-Gyeong Yun and Man Gu Kang
Convergence Components & Materials Research Lab., Electronics and Telecommunications Research Institute (ETRI), Daejeon, Republic of Korea

Byeong-Soo Bae
Lab. of Optical Materials and Coating (LOMC), Dep. of Materials Science and Eng. KAIST, Daejeon, Republic of Korea

Yongseok Jun
Interdisciplinary School of Green Energy, Ulsan National Institute of Science, Ulsan, Republic of Korea

Daniele Colonna, Daniele D'Ercole, Girolamo Mincuzzi, Thomas M. Brown, Andrea Reale, Aldo Di Carlo and Lorenzo Dominici
Centre for Hybrid and Organic Solar Energy Centre (CHOSE), Dept. of Electronic Eng., Tor Vergata University of Rome, Roma, Italy

Francesco Michelotti and Lorenzo Dominici
Molecular Photonics Laboratory, Dept. of Basic and Applied Physics for Eng., SAPIENZA University of Rome, Roma, Italy

Riccardo Riccitelli
DYERS srl, Roma, Italy

King-Chuen Lin and Chun-Li Chang
Department of Chemistry, National Taiwan University, Taipei 106, Institute of Atomic and Molecular Sciences, Academia Sinica, Taipei 106, Taiwan

Xiang-Dong Gao, Cai-Lu Wang, Xiao-Yan Gan and Xiao-Min Li
State Key Lab of High Performance Ceramics and Superfine Microstructures, Shanghai, Institute of Ceramics, Chinese Academy of Sciences, Shanghai, P. R. China

Matthew J. Griffith and Attila J. Mozer
ARC Centre of Excellence for Electromaterials Science and Intelligent Polymer Research Institute, University of Wollongong, Squires Way, Fairy Meadow, NSW, Australia

William A. Vallejo L., Cesar A. Quiñones S. and Johann A. Hernandez S.
Universidad Nacional de Colombia, Universidad de Cartagena, Universidad Distrital F.J.D.C, Bogotá, Colombia

Masaya Chigane, Mitsuru Watanabe and Tsutomu Shinagawa
Osaka Municipal Technical Research Institute, Japan

Qiquan Qiao
South Dakota State University, United States

A.P. Uthirakumar
Nanoscience Centre for Optoelectronics and Energy Devices, Sona College of Technology, Salem, Tamilnadu, India

Elias Stathatos
Technological-Educational Institute of Patras, Electrical Engineering Department, Patras, Greece

Printed in the USA
CPSIA information can be obtained
at www.ICGtesting.com
JSHW011422221024
72173JS00004B/643